Networking *and* Integration *of* Facilities Automation Systems

Networking *and* Integration *of* Facilities Automation Systems

Viktor Boed

with contributions from:
Ira Goldschmidt, Robert Hobbs, John J. McGowan,
Roberto Meinrath, and Frantisek Zezulka

CRC Press
Taylor & Francis Group
Boca Raton London New York

CRC Press is an imprint of the
Taylor & Francis Group, an **informa** business

Contact Editor: Cindy Carelli
Project Editor: Maggie Mogck
Cover Design: Dawn Boyd

CRC Press
Taylor & Francis Group
6000 Broken Sound Parkway NW, Suite 300
Boca Raton, FL 33487-2742

© 2000 by Taylor & Francis Group, LLC
CRC Press is an imprint of Taylor & Francis Group, an Informa business

First issued in paperback 2019

No claim to original U.S. Government works

ISBN 13: 978-0-367-45558-3 (pbk)
ISBN 13: 978-0-8493-0699-0 (hbk)

Visit the Taylor & Francis Web site at
http://www.taylorandfrancis.com

and the CRC Press Web site at
http://www.crcpress.com

Library of Congress Card Number 99-15538

Library of Congress Cataloging-in-Publication Data

Boed, Viktor.
      Networking and integration of facilities automation systems / Viktor Boed ; with contributions
   from Ira Goldschmidt...[et al.].
            p.   cm.
      Includes bibliographical references and index.
      ISBN 0-8493-0699-X (alk. paper)
      1. Commercial buildings--Automation.   2. Intelligent buildings.   3. Digital control systems.
   I. Goldschmidt, Ira.   II. Title.
   TH6012.B66   1999
   696—dc21

                                                                              99-15538
                                                                              CIP

Preface

Facilities increasingly rely on computerized systems for optimization of building systems operation and reduction of cost for building maintenance and management. Systems networking and their integration is becoming a new discipline for facilities managers and engineers as more and more computerized systems are being implemented.

Automation systems for facilities are in two categories:

- **Real-time** controls and automation systems, such as direct digital control (DDC) and automation systems, power plant controls and automation systems, industrial control systems, online utilities metering systems, and other specific systems
- **Data processing systems**, such as maintenance management systems, production control systems, various office automation systems, engineering systems, drafting and project management systems, and other specific systems

The above systems are designed to aid facilities managers and owners to reduce operating, maintenance, and management costs.

Historical development of discrete systems with vendor proprietary protocols resulted in having a number of systems and isles of automation in every facility. Such systems could not provide adequate and effective tools for facilities managers to optimize their operation and minimize operating cost. Facilities automation systems must share information with each other and must be designed with interfaces that allow interoperability over facilities networks.

Up to the mid 1980s, most automation systems were offered with proprietary communications to their own "front-end" computers. This landscape is changing by increasing demand for information exchange and systems interoperability. Most automation systems on the market provide limited interface options with systems within their own category (for example, DDC systems). However, most systems lack capability to interface outside of their own category (for example, DDC systems to power plant control and automation systems).

When considering systems interoperability, one has to define interface requirements as:

- Interfaces of different vendor's systems of the same category, i.e., different vendor's DDC systems. Most DDC systems have several levels of hierarchy with varying communications requirements (i.e., communication on the facility level, building level, application specific level). Therefore, interface requirements should be defined at each specific level.
- Networking of real-time systems of different categories (i.e., interface of building automation systems to power plant controls and automation systems), interface of energy monitoring systems to production control systems, and so on
- Interfaces of real-time automation systems to database type systems (i.e., DDC system to maintenance management systems)

Depending on the complexity of the facility, there may be other requirements to enhance information exchange, optimize production, minimize energy consumption, schedule maintenance, manage material and manpower, and other requirements.

Site-specific needs should define the design requirements for networking. Another defining factor should be the utilization of data. Some applications will require high speed data exchange between two computers for control, monitoring, or data processing; other interfaces may be designed to provide import and display of data on PC work stations. A client from a PC window environment can access and retrieve the data by opening a "window" to different network computers or servers in a similar fashion as we routinely access the Internet, E-mail, or other programs from our desktop personal computers.

To have a successful implementation of systems networking, engineers and facilities managers have to understand the basics related to networking. Only then can they make informed decisions as to definition and design of networking requirements, and provide continuous support for installed networks.

Most facilities engineers and managers are trained in engineering disciplines related to building and HVAC industry or in controls and automation systems. Communications and networking are new to most of us.

Most communications-related publications focus on individual protocols. Besides aiding the end users in understanding the "buzzwords" of the communications industry, the intention of this book is to discuss the basics of networking and selected communication protocols and drivers. The book also explores interface options between systems of different categories. While such interfaces are outside of the "real-time controls" provided by automation systems, the information they provide has great value for facilities engineers, administrators, and managers. Their greatest value is in providing real-time data to clients on the network and in compiling data for online analysis, systems performance, and cost optimization. Networks providing global information online become invaluable tools for engineers and managers in their daily routines, in evaluation of systems performance, resource management, and cost allocation.

Networks, protocols, and drivers discussed in the book are commercially available in the market. Performances of individual network components, protocols, or drivers are not evaluated in the book. It is up to the end users or systems integrators to evaluate the characteristics of individual network components and to match them with the unique needs of the facility for optimum performance.

The case study presented in the last chapter is based on a large-scale integration effort spanning several years. The network is implemented from commercially available components. The chapter also provides an account of the benefits such large scale integration can provide the end user with regarding optimization of systems performance and consequent reduction of operating cost.

To introduce other viewpoints to the subject of systems interoperability, I've asked several people for their contributions. Some of the contributions are written specifically for this book, some are reprinted from magazines with the authors' permissions. The intention of this approach is to present the related subjects from different angles for the readers. The contributions are reprinted in their original versions, without editing by the author of this book.

Since each facility is unique, systems integration is like putting together a large puzzle. It would be a mistake to wait for a "miracle," for a generic "protocol" which would provide an answer to all our integration needs. Protocols and drivers are constantly evolving; the ones discussed in the book address only a fraction of facilities automation needs.

In an ever-changing world of computers, automation systems, communications, and the desire of facilities for cost reduction, systems integration is still a challenging endeavor. Only educated end users and integrators who understand the needs of facilities and their automation requirements and are familiar with commercially available network components can rise to the challenge and succeed with facilities integration. The real challenge is to preserve the owner's investments into systems and networking by creating an integrated network that can grow with the evolving technology, without obsolescence of its major components or without major changes of its configuration.

The Author

Viktor Boed, CEM (Certified Energy Manager), has a graduate degree in electrical engineering from the Technical University Brno in the Czech Republic. As a consulting engineer he has designed building automation systems in Europe and the U.S. and taught courses in electrical engineering at the Technical University in Brno as well as in Cairo, Egypt. Boed came to the U.S. in 1979 and worked as a product manager and senior research engineer for Johnson Controls, Inc. in Milwaukee, WI. He joined Yale University in 1983 as Manager of Building Automation and became Manager of Plant Engineering in 1989, where he is involved today in design and implementation of automation systems for buildings, power plants, maintenance management systems, utility metering systems, and a facilities real-time communications network. The Plant Engineering Division also is involved in implementation of energy conservation projects, and engineering review of capital projects for the university.

Boed is an active member of the Association of Energy Engineers and started the Connecticut AEE chapter. He is the recipient of the Energy Manager of the Year and the regional Energy Engineer of the Year awards, and is a frequent speaker at World Energy Engineering Congresses and other professional conferences. He also is a member of the American Society of Heating, Refrigerating, and Air Conditioning Engineers (ASHRAE) and was on the ASHRAE SPC 135 committee that developed the ANSI/ASHRAE Standard 135/1995: BACnet, a data committee protocol for building automation and control networks.

Acknowledgments

I wish to thank John McGowan, Ira Goldsmidt, Prof. Frantisek Zezulka, and Bob Hobbs, for their contribution to this book. I also wish to thank Bill Payne, Editor in Chief of Strategic Planning for Energy and the Environment, for putting me in contact with some of the authors. Special thanks to Roberto Meinrath, a friend and boss, for his contribution, and for editing the manuscript just as he did for my previous two books.

Contributors

Viktor Boed
Manager of Plant Engineering
Yale University
New Haven, CT

Ira Goldschmidt
Senior Engineer
RNA Design
Denver, CO

Robert Hobbs
Sr. Systems Engineer
Plant Engineering
Yale University
New Haven, CT

John J. McGowan
Vice President of Marketing
Hydroscope Group, Inc.
Albuquerque, NM

Roberto Meinrath
Deputy Director of Facilities
Yale University
New Haven, CT

Frantisek Zezulka
Associate Professor
Department of Controls and Automation
Technical University Brno
Czech Republic

Contents

For Hana

1 Integration of Facilities Computerized Systems

Viktor Boed

CONTENTS

SYSTEMS FOR FACILITIES AUTOMATION

Management of facilities — from single office buildings through rental properties to large multibuilding facilities — has become a complex task involving multitudes of disciplines. With increased awareness for efficient building operation, building owners and operators are looking for methods to reduce the cost of building operation.

Cost optimization of facilities is associated with such disciplines as:

- Optimum space utilization
- Reduction of administrative overhead
- Reduction of engineering and development cost
- Optimization of energy cost (both production and usage)
- Reduction of maintenance cost
- Optimization of building systems operation
- Automation of power plant operation
- Automation of production
- Facilities data management, analysis of data provided by different facilities systems interfaced on the same network
- And other categories

To effectively address the complex task of modern facilities management, facilities managers increasingly rely on computerized systems designed to optimize building operations, maintenance, and management. Industries associated with "buildings" responded to the challenge of automation by developing computerized systems aimed to automate building operation, maintenance, and management. Since the above-listed disciplines are not interrelated, their associated computerized systems have very few, if any, similarities. For example, maintenance management systems (MMS) are being developed and marketed by systems houses with expertise in computerization of work scheduling, maintenance planning, inventory control, and accounting. Building automation and control systems (BACS, BAS) or energy management and control systems (EMCS, EMS) are being developed and marketed by controls and automation vendors. The two entities belong to different "industries," yet they serve the same market segment — *automation of facilities operation*. Despite their dissimilarities, they were designed and are implemented to optimize their respective areas of building operation. The two systems could complement each other by automating information transfer from one system to another.

In the above example, an alarm reported by a BAS (acknowledged and checked out by an operator) can be transferred over the network to the MMS for work scheduling, trade assignment, material request, and allocation of time to complete the task, followed up with work planning, scheduling, and accounting activities. Such interaction of the two systems assures faster response to the problem, reduced time for work completion, reduced work load for the operator, reduced downtime of the HVAC system, and reduced operating and maintenance (O&M) cost. The most commonly utilized automation systems by modern facilities fall into the following categories:

- Administrative systems
- Accounting systems
- Space-utilization systems
- Inventory control systems
- Maintenance management systems
- Engineering systems (i.e., AutoCad)
- Building automation and control systems (i.e., DDC)
- Specific production or laboratory automation systems
- Fire, security, access control systems
- Lighting control systems
- Elevator control systems
- Utilities or power plant automation systems
- Utilities metering system
- Other facilities specific systems

Due to historical development of systems within each of the above categories, even "well" automated facilities operate multitudes of stand-alone systems, vendor proprietary systems, and discrete local area networks (LANs).

Considering the characteristics of computerized automation systems for facilities, they are either:

- **Real-time systems** — these acquire data from associated field hardware points at pre-defined scan rate (i.e., real-time controls and automation systems) or
- **Data management systems** — the data is acquired by manual entry from the keyboard (i.e., accounting systems), by file transfer (i.e., the parts request is transferred from the MMS to the inventory control system), by scanning (i.e., bar code reading for the inventory control system), or by combination of the above entries

The difference between the two systems is in scheduling (scanning) and update of data acquisition. Real-time systems such as distributed digital control (DDC) systems must operate with "realistic" scan rates to assure accuracy of their control action. For example, accuracy of a control loop depends, besides other parameters, on the scan rate — on the elapsed time between acquisition of analog data from a field sensor and the time the command was issued to the control valve. There, a fast scan rate is essential. Frequency of data acquisition of data management systems depends on the frequency of data inputs and outputs.

SYSTEMS INTEGRATION

Only thoughtful approach to systems integration, thorough understanding of the needs of the facility, and advanced planning can bring the desired results for systems integration.

For new facilities, it means specifying requirements for systems integration early in the program development phase, studies, and project justifications, and working with a systems integrator throughout all phases of the project. Such early involvement will result in adherence to goals for systems interoperability, and inclusion of interoperability issues into the design and development phases of the project. This will then assure selection and implementation of systems, which will communicate over the proposed networks utilizing standard protocols or drivers.

For existing facilities, the task of systems integration is more complex. Since very few existing facilities were developed with systems integration in mind, the goal for existing facilities is to determine the needs for systems integration and match them with networking capabilities of existing systems. The most realistic approach for integration of existing systems should be based on initial surveys of the facility.

Due to technological differences of systems implemented over long time periods (a year can make a big difference when it comes to systems integration), integration of existing systems is a complex and tedious task. Such a task should be accomplished by adopting a long-term view for facilities and systems automation. Upon identifying the needs and integration capabilities of existing systems, an overall long- and short-range plan should be developed for integration.

On one hand, adherence to the integration plan is essential for successful systems implementation. On the other hand, due to ever-evolving systems development and the nature of the computer and the automation and communications industry, integrators should be prepared to be flexible in their approach, and implement state-of-the-art systems as they become available. Integration of existing systems in large facilities could be a long-term process spread out over several years.

The greatest challenge for facilities managers and system integrators is to make decisions which would assure implementation of new technology and new systems and approaches into networks and systems already in existence, without obsolescence of existing systems. This is an important consideration, since building owners have invested considerable amounts of money and time into the development of existing systems and networks (Figure 1.1A).

Utilizing advanced networking technologies, computerized facilities systems, and presentation of data on commonly used hardware and software products (i.e., PCs, DOS, Windows, spreadsheets), system integrators can achieve a high level of integration which will be accepted (and used) by the clients on the network. This is especially important for "facilities management" type of data, imported and formatted (using popular pieces of software) for the network users for presentation on their workstations (i.e., online import of real-time data into existing spreadsheets) (Figure 1.1B).

To avoid pitfalls, it should be emphasized that systems integration and information management over a network (Figure 1.2) is an entirely different discipline from, let us say, systems engineering of a BAS or implementation of a MMS. System integrators dealing with integration of facilities automation systems should focus on two major tasks:

1. Design of the most suitable network and interface protocols for the facility and its automation systems.
2. Definition of information flow and its management.

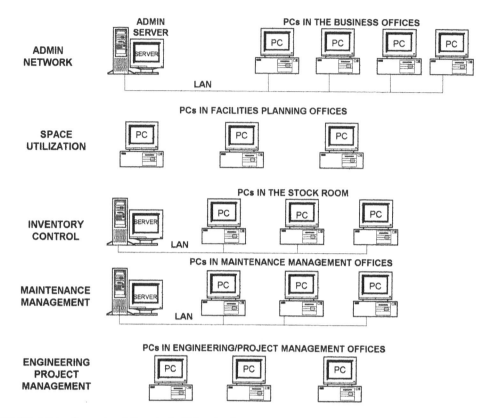

FIGURE 1.1A   Example of discrete facilities database systems.

The first item is an engineering task, which should never be undertaken without thorough understanding of the needs of the network users. There is an analogy between a systems integrator's understanding the network users' needs and a building automation systems engineer's understanding of the building mechanical systems (HVAC systems).

Based on their understanding of the needs of human users and the connected computers and controllers on the network, system integrators should define the characteristics of the information to be transferred over the network. This definition should involve information to be transferred, points to be mapped over the network, sizing of data files, definition of transmission speed or rate, data presentation on the user PCs (i.e., tables, trend graphs, color graphics, spreadsheets), and other information management and data presentation related issues. Optimization of information flow is essential. Although it is possible to transfer gigabytes of information at high speed over a high speed network, in practice, *not all users of the network need all the information available on the network all the time!*

CONDITIONS FOR SYSTEMS INTEROPERABILITY

For computerized systems to be able to communicate with each other, the following compatibility issues should be resolved:

1. Systems should have similar or compatible **operating systems** (DOS, OS/2, UNIX, etc.) with similar networking capabilities.
2. Systems should be on the same or compatible **networks** (Ethernet, ARCNET, etc.), or should utilize network interface devices, such as convertors, gateways, etc., to resolve network incompatibility issues.

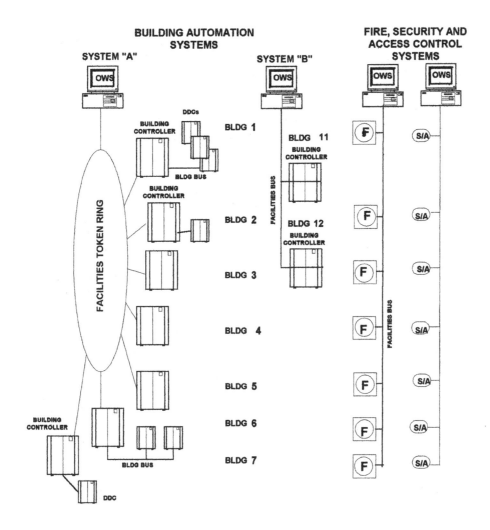

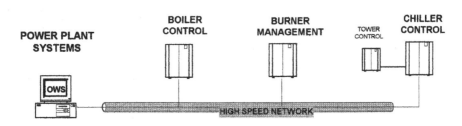

FIGURE 1.1B Example of discrete real-time facilities automation systems.

3. Systems should have compatible **application program interfaces** (NetBIOS, Sockets, etc.).
4. Systems should have the same or compatible **communication protocols** *or* **drivers** (OSI, TCP/IP, CAB, BACnet, Modbus, etc.).
5. Systems should have their **application software** designed to accommodate the information imported from other systems over the network (mapover points, using available pieces of software, such as Net DDE, Bolt_On, etc.).

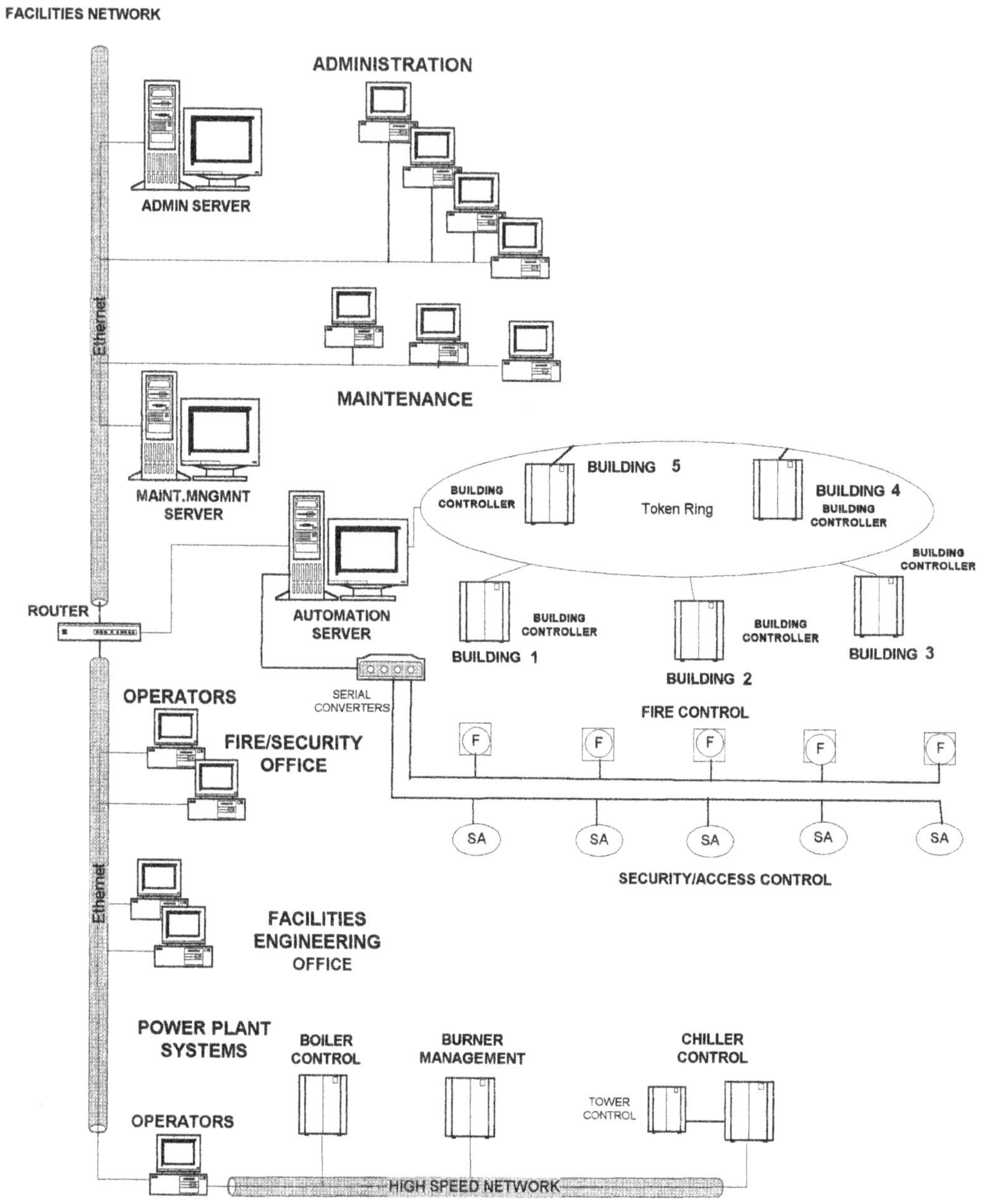

FIGURE 1.2   Example of an integrated facilities network.

The following table is an example of off-the-shelf systems networking components matched for compatibility.

Table 1.1   Network Compatibility

Hardware	Operating Systems	Communications Protocols	Networks	Application Program Interface
IBM-PC	MS-DOS	BACnet	ARCNET	NetBIOS
SUN	UNIX	TCP/IP	Ethernet	Sockets

Facilities managers attempting integration should understand the above compatibility requirements and should rely on system integrators to avoid costly mistakes.

Each of the previously mentioned facilities automation system categories represents a segment of the industry — an independent engineering discipline. There is a high probability, for instance, that vendors and engineers designing and implementing a building automation system *will not have* in-depth knowledge of integration of their system into systems from other categories (let us say, to MMS), and vice versa. Besides, system vendors are paid to sell, engineer, and implement their own systems. Even with the best of intentions, two system vendors may propose systems — perhaps the best in their own categories — which could not be integrated on a given facilities network without major and expensive modifications.

Systems integration is complicated by the fact that there are numerous systems on the market within each category. Such vendor-specific systems may run on different hardware platforms; utilize different operating systems, communication protocols, or communicate on different networks; and may have different application program interfaces (APIs). Since there is more than one vendor-specific system in each category — each with several networking options — systems integration is a task of matching the individual system's options to the integration requirements of the given facility. The process is sort of "pick and chose" from standard systems on the market, with the goal to fit all pieces together for systems **interoperability**.

Facilities engineers and systems integrators should define networking needs in two ways:

1. Networking within each group (i.e., networking of building automation systems).
2. Networking between groups (i.e., interface of BAS to power plant automation systems, to MMS, etc.).

For example, BAS protocols and standards, such as the Canadian automated building (CAB) or the ASHRAE BACnet protocols, standardize the networking requirements for BAS. However, they do not address fully networking issues outside of the BAS category.

It should be said that in the absence of governing standards, simple matching of standard systems does not work all the time. For many existing facilities, customization of standard systems, or at least of their interfaces, is unavoidable. The good news about custom interfaces is that systems modifications are — in most instances — related to systems interfaces only, and they do not affect the application software of, let us say, a BAS. Development of custom interfaces does not negate the system's warranty or continuous vendor's support, if such modifications are discussed with, and agreed upon by, the involved system vendors. The bad news about custom interfaces is that, in most instances, they are written by a system integrator for a specific software of a specific system residing on the network (i.e., Rev. level 1.1 of a BAS vendor A software).

Assume that a facilities manager commissions a system integrator to write interface software (protocol or driver) for a BAS "A" with Rev. 3.1. application software to communicate with another system "B," which has a Rev. 5.0 application software, on a site specific network. Two years later, BAS vendor "B" releases a new, Rev. 6.0 software upgrade, and offers the upgrade to the customer. Prior to purchasing the upgrade, the facilities engineer responsible for the system upgrade should find out whether the new software is compatible with the existing interface software written for the original revision levels.

There is a good chance that the newly released software (Rev. 6.0) would not be compatible with the custom-written interface software (Rev. 5.0), unless the interface software is a so-called "standard" or "industry standard" protocol or driver accepted by most automation systems vendors.

Therefore, prior to implementing a newly released automation software for already networked systems using a custom-written protocol, the network compatibility issues should be verified to assure systems interoperability.

Considering the life cycle of automation systems and the rapid development of communication interfaces, it is advantageous to implement interfaces, which were tested by vendors, installed at

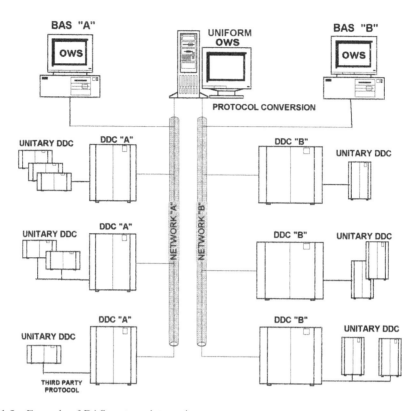

FIGURE 1.3 Example of BAS systems integration.

other customer sites, and are supported by the vendor's local office. Automation system vendors publish a list of supported drivers or interface protocols for their systems. In cases where the customization is unavoidable, the customer should fully understand the related issues, advantages, and limitations of customized interfaces.

Facilities automation systems are constantly evolving, due to new development of software, hardware, and communication options. System integrators have to coordinate, for each new system, compatibility of **hardware**, **operating systems**, **application program interfaces**, **communication protocols**, and **networks** to maintain interoperability of systems connected to the network.

Interoperability of systems is not only an engineering but also an information management challenge. System integrators along with facilities engineers have to determine the *degree of systems interoperability for a given facility*. The task is similar to determining the *degree of automation* by the systems engineer for BAS applications.

NETWORKING BUILDING AUTOMATION SYSTEMS

To demonstrate the decision-making process related to systems interoperability, let us look at the networking requirements in an example in Figure 1.3. To keep the example simple, the systems are all in the same (BAS) category. The goal is to meet the operating requirements of the owner, while providing a reliable and economical solution.

A group of buildings is controlled by BAS systems from vendor "A". Another group of buildings is controlled by BAS systems from vendor "B". There is a chiller in building A3 with its own controller. The computer-based chiller controller is a so-called third party (or application-specific controller) provided by the chiller manufacturer. Let's call it controller "C". All laboratory buildings are controlled by BAS systems "B". The related fume hood controllers are from vendor "D".

Let us look at facilities requirements and operating practices:

1. The facility in the example has central dispatching, with two sets of operating work-stations (OWSs): one set for BAS system "A", another set for BAS system "B". The project scope requires integration of the two BAS systems into one common OWS, with uniform data presentation for the operators.
2. Building automation systems "A" and "B" are dedicated to certain buildings; there is no combination of systems "A" and "B" in any one building (a prior decision made by the management of the facilities to keep buildings vendor specific). Consequently, mapping over points between systems "A" and "B" for control purposes is not required.
3. Specific chiller points are to be remotely monitored by the central dispatching; some points are to be mapped over to BAS system "A"; chiller safety-related points have to be hardwired to the BAS controller "A3", as per the following requirements:
   a. Defined number of points from chiller controller "C" have to be mapped over to BAS "A" via industry standard protocol or driver
   b. "Safety"-related points have to be hardwired from chiller controller "C" to the BAS controller "A3"
   c. Since chiller controller "C" is microprocessor based, internal chiller control points (i.e., control loop parameters) do not have to be mapped over to the BAS system
4. Fume hood controller "D1" controls a critical laboratory area. All points from that area have to be monitored by central dispatching. In the other laboratories, controlled by BAS controllers "B2" and "B3", the researchers themselves are responsible for laboratory safety and fume hood operation (fume hood controllers "D2" and "D3"). Central dis-patching does not monitor operation of the fume hoods in these areas. Consequently, essential points from fume hood controller "D1" have to be mapped over to BAS "B". Points from fume hood controllers "D2" and "D3", related to volumetric control (supply and exhaust CFM), night setback, and other "building" functions should be hardwired to laboratory BAS controllers "B2" and "B3".

The above example demonstrates an approach to interoperability by defining site-specific requirements for the systems integrator. Division of responsibilities (especially in areas controlled by system "B") is between the laboratory users (in areas controlled by "B2" and "B3") and the facilities management (in area "B1"). The decision related to hardwired vs. points mapped over on the network should be based on operational as well as economic considerations. For example, points, which can be locally accessed and are locally displayed for operations and maintenance (i.e., chiller tuning parameters or face velocity of individual fume hoods), do not need to be mapped over the network. Points that are critical for safety and operation (i.e., chill water flow, pressure, or volumes of laboratory exhaust and supply air) should be hardwired between the building controllers and third-party controllers (i.e., between chiller controller "C" and BAS controller "A3", or fume hood controller "D2" and VAV controller "B2").

The above approach is completely different from a "standard protocol approach," which assumes that all systems have a common (same) protocol, and all information is available on a common network for all users of the network (see Figure 1.4).

It should be pointed out that at present there is no such protocol in existence. Even if there were a protocol accepted by **"all"** vendors providing systems for the building industry (real-time and data management systems), information management and information optimization (who needs what information and what to do with it) should be defined to optimize the available facilities resources. Availability of information should be in conjunction with responsibility for information management (i.e., who is responding to a low-face velocity alarm from a laboratory fume hood during operating and during off-hours). In the above example, mapping overall fume hood points over the network and providing the information to the system operators puts responsibility (and liability) for fume

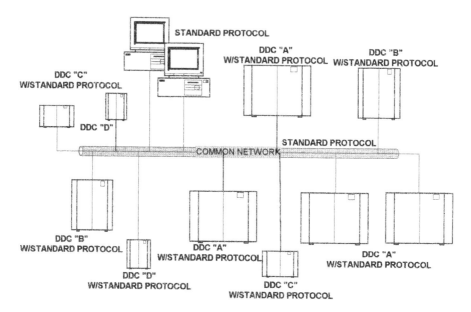

FIGURE 1.4 Systems integration with common communications protocol.

hood operations and related laboratory safety on the facilities operators. Such additional responsibility would be a substantial increase of work load for operators of large facilities with hundreds of fume hoods, and a substantial increase of their existing responsibilities. Researchers working in the laboratories have more understanding of the operating practices and laboratory safety, and can react to the changing (local) conditions more readily and with a higher degree of knowledge and responsibility than the facility operators involved in remote monitoring of multiple buildings.

NETWORKING FACILITIES AUTOMATION SYSTEMS INTO FACILITIES INFORMATION AND DATA MANAGEMENT SYSTEM (FIDMS)

Interoperability of systems of different categories on a facilities network is a new challenge for most facilities and systems engineers. The "facilities information and data management" concept is based on the desire of facilities managers and engineers to have access to data available on systems connected to the network. The distributed data are then used by various departments of the facilities, such as:

- Facilities engineering to evaluate real-time data, systems performances, energy consumption and savings, building energy profiles, troubleshooting of connected systems, provision of measured data for future design and development, engineering reporting, and other
- Production engineering and management to provide production planning, performance analysis, systems troubleshooting, production reporting, and other
- Utilities department to provide measurements of utilities, production, import, export, accurate measuring of consumption in individual buildings or users, utilities billing, reporting to state and federal authorities, and other
- Maintenance department providing inventory control, generation of work orders, work planning, accounting, time reconciliation, analysis of maintenance records, and costs for individual units, systems, building, and for other O&M-related tasks
- Facilities management to analyze costs associated with building operation and maintenance; management report generation; generation of reports for local, state, and federal authorities; and other management tasks

- Building users and customers, by providing them with information associated with their environmental control systems, usage, and allocation of energy, energy savings opportunities, information on scheduled maintenance, and other building-related issues

Custom programs providing individual users with the desired information can extend the above-outlined benefits. Due to networking, and the **facilities information and data management** concept, the information is easily obtainable by the network users and, therefore, the information is **being used daily**. This flow of information frees the facilities managers from many of the administrative burdens associated with obtaining information from the individual systems of the facility, analyzing the data, and generating the reports. FIDMS places the information on the network and makes it available (online) to users for analysis and decision making.

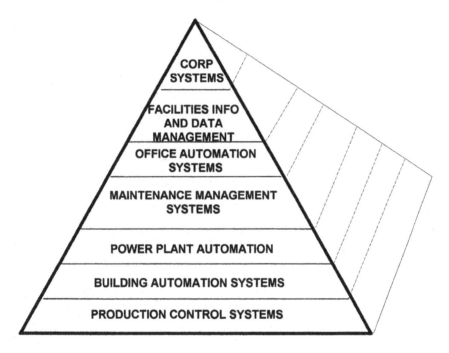

FIGURE 1.5   Facilities information and data management concept.

The concept is sort of a pyramid (Figure 1.5) with the information and data management system (IDMS) on the top, interfacing to corporate management systems, and on the lower levels to office automation systems, MMS, utilities metering systems, energy and power plant systems, BAS, and production control systems.

The FIDM concept expands the challenge for systems integration from integration within one category (i.e., BASs) to integration of all or most of the facilities automation systems into one FIDM system. Integrators of FIDM systems must pay increased attention to design of:

- Communication networks and their architecture
- Communication protocols to be utilized for individual systems and networks
- Data presentation to users of the system
- Support of the information on the network
- Network management, security, reliability, and integrity

The above items are new to most systems engineers practicing applications engineering of BAS, and certainly new to most facilities managers and engineers. Due to segmentation of automation systems utilized by facilities, it is difficult to find integrators familiar with "all" facilities automation systems, networks, and protocols. In most instances, facilities have to rely on expertise of individual automation vendors' application engineers, and find capable **system integrators** to define the facilities' needs and develop the desired integration.

DEFINITION OF COMMUNICATION NETWORKS AND THEIR ARCHITECTURE

Networks for each level of systems hierarchy are determined by individual systems at that particular level. For example, BAS systems may use different communication at each level of the system architecture (serial communication on the controller level, ARCNET on the building level, Ethernet on the facilities level). Network topology will depend on the layout of the facility, location of individual nodes (controllers, user PCs), available communications media (fiber optics, telephone network, dedicated wiring). Consequently, use of network devices will also depend on the individual systems, distances, number of nodes on the networks, etc. Communication speed is another important network parameter, which will determine the performance of the network. This will have to be determined for each individual system and associated network.

The diagram in Figure 1.6 is an example of the frequency of data transfer pertinent to individual systems residing on the network. While production control and building automation systems operate at fast scan rates, other system updates are at longer time intervals (in minutes) or require updates per demand (scheduled or per request). This is an important consideration for efficient design of networks and cost optimization.

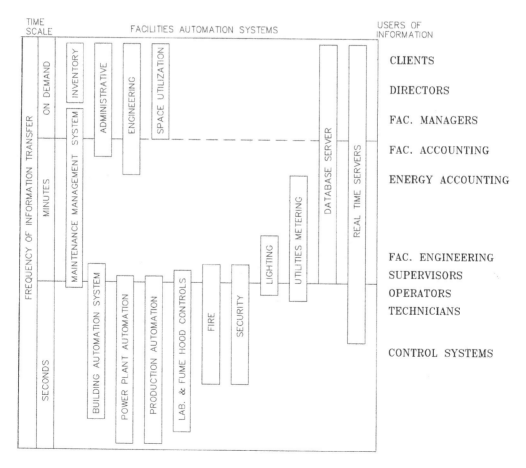

FIGURE 1.6 Frequency of information transfer for individual systems.

Definition of Communication Protocols and Drivers

There are literally hundreds of proprietary, industry standard, or standard protocols and drivers on the market. Every systems vendor provides a list of available protocols and drivers for their own systems. System integrators are in the business of writing custom drivers for systems integration, or integrating systems using commercially available protocols and drivers. The challenge is to define the most appropriate driver or protocol for each system to be integrated.

Facilities should standardize on proven drivers and protocols and implement the ones most suitable for the facility. Otherwise, the facility could end up with a maze of protocols and drivers that would be costly to implement and difficult and costly to support.

Data Presentation to Users on the Network

Individual systems on the network communicate on common or dedicated networks which, in most cases, provide data transmission from real time as well as data management systems. Two related issues are very important — information management and information presentation.

Information Management

Systems integrators should interview the facilities staff and determine:

- What information is required on individual nodes of the network (i.e., the BAS system operator needs information on chilled water flow, pressures, and temperatures from the utilities metering system. However, he does not need information on cooling tower temperatures of the generating plant).
- How to process the received information (i.e., a power plant manager would not know what to do with a high temperature alarm of a particular room controlled by a BAS, even if he had received it on his PC).

Note: if we do not have positive answers to the above questions, the information is probably not needed for that particular user.

- Network nodes or systems should be autonomous in information processing. Information transfer among nodes should be minimized (i.e., an ultrasonic chilled water flow meter should convert the ultrasonic signal to flow [in fps or GPM] and calculate the connected chilled water consumption [tonnage or BTU/h], if the supply and return temperature sensors are connected to the meter. This integration on the low level will increase the data accuracy and will reduce the network traffic, when compared to scanning each field point individually and executing the relevant calculations by a network server or dedicated computer).
- The frequency of data transfer (see example in Figure 1.6).
- Format of data files to match the recipient's file structure (i.e., coma-separated variables or CSV files for data imported into spreadsheets; DXF files for import of graphic data into AutoSketch or AutoCad; etc.).
- Size of files to store information received over the network (i.e., file sizes for daily electrical reports reported at 15-min demand intervals should hold at least 1 month's worth of data).
- Archival of databases and historical information on dedicated computers on the network (i.e., BAS database resides on one of the BAS computers; building utilities metering historical data reside on a dedicated facilities server; power plant daily report archive resides on a dedicated power plant computer, etc.).
- Categorization of information needed for systems engineering, operation, maintenance, and management (i.e., the building maintenance department needs information from the

BAS, MMS, and building utilities metering; it does not need information on utilities billing or power plant operation).
- What information is to be transmitted over the network to other systems (i.e., from BAS to maintenance management system, from BAS to the FIDMS).

The above generic information is listed as a guideline. Each facility has its own particular systems and specific requirements related to information transfer. Systems engineers should interview the customers connected to the network and determine the individual department's (or node's) requirements during the early stages of system design.

Information Presentation

There should be a clear distinction between **systems specific information** (i.e., BAS information presented on the BAS-OWS; MMS information presented for the maintenance personnel; project management data presented to project managers, etc.), and **engineering and management information** (i.e., highest kW demand per day of a particular building, total kWh per month of a building; reports on number of work orders issued for a particular air-handling unit per month; total cost of renovations and repair for a building, kWh usage per square foot of a specific building, etc.).

Systems-specific information is usually presented on the displays of controllers, system-specific operating workstations (OWS), or personal computers (PCs) in a systems-specific format. The display format is often vendor specific (i.e., a BAS vendor's person machine interface — PMI, MMS screens of a particular MMS vendor, etc.).

Management information, such as management reports, spreadsheets, graphics, trend graphs, etc., compiled from individual systems, is distributed over the network and presented on user PCs. The data format is usually determined by the PMI screens of the FIDMS server and by the systems engineer developing the interface screens. No matter what server is used, the data should be easily accessible, compiled, and presented in a familiar format, such as in word processing (Word), spreadsheets (Lotus, Excel, dBASE, etc.), graphic software (Harvard Graphics), drafting software (AutoCad), trend graphs, and other formats familiar to the network users.

Depending on the type of the user, the FIDM data should be presented either on a dedicated PC (i.e., power plant and distribution system's data presented on a dedicated PC for the power plant manager), or on office PCs along with other PC features and programs. If the data are presented on an office PC, the systems engineers should set up the user PCs so that access to facilities data would be as simple as access to any other software or program available on the user's PC. In a PC window environment, different icons can represent different groups of data (i.e., building electrical metering system). By clicking on the icon, the user should access a focus window to the electrical metering system. Depending on the access level of the user, he or she can then view the electrical metering system from overall summaries and reports, to trend graphs, or with high access level passwords, the data residing in individual meter registers.

There are many forms of data presentation. The key to successful implementation is the user's familiarity with the particular interface software. Such familiarity is based on prior experience with that piece of software or on comprehensive training. Many otherwise good systems fail, due to lack of vendor or systems integrator's consideration for the users. Remember, the users are a powerful group of people when it comes to systems acceptance, or its failure.

SUPPORT OF THE INFORMATION ON THE NETWORK

Information put on the network has to be credible and presented in a format the users can understand. Otherwise, the organization that puts such information on the network will be overwhelmed with questions from the users. Since the nature of the information is dynamic, most information has to be updated periodically and supported by the staff responsible for the information. For example, daily energy consumption for each building or department can be reported in the form of a table and trend graph. Such information is useful for the business or department managers to assess their energy cost and peak demand and to encourage energy conservation. The information available on the network has to be accurate, consistent, and credible, and has to be supported by the utilities department. In a multibuilding facility, there will be (always some) customers who will need clarification, explanation, or just information related to the data presented.

To achieve the desired advantages and productivity increases due to systems integration, customer support should be provided internally by one of the departments of facilities management rather than by an outside organization.

NETWORK MANAGEMENT, NETWORK SECURITY, RELIABILITY, AND INTEGRITY

Large networks consisting of several different systems, servers, and network devices are complex and require engineering and management support. Large networks should have a dedicated support and network management. Network support should be considered at the earliest stages of planning, since it usually means increased operating expense for the owner. Network support can be provided by in-house staff or subcontracted to systems houses.

Network security of large networks is a delicate balance between security measures and convenience of the users to access data over the network. Every network has to be protected from access by outside, unauthorized users. In addition, individual systems have to be protected from unauthorized network users who intentionally or unintentionally can cause damage to the connected systems, data, or equipment. Unauthorized access to real-time systems could turn into a safety nightmare, considering the large equipment controlled by such systems. Real-time systems (such as power plant control and automation systems) require a higher level of security than just user password or assigned addresses in the network devices. Implementation of an appropriate "fire wall" protects the connected systems without inconveniencing the network users.

Reliability and integrity of the network is another issue to be considered from the early stages of design. Besides design considerations, the network integrity has to be protected throughout the life of the network — its maintenance, repairs, and upgrades. Many networks utilizing existing data-grade telephone circuits, initially in good working condition, deteriorate over time, due to constant modifications of the telephone circuitry and unintentional interference with the circuits dedicated for network communications. To prevent such occurrences, separation of the network circuitry on dedicated terminal strips and their proper labeling can prevent many hours of troubleshooting of network-related problems.

2 Basics of Network Communications

Viktor Boed

CONTENTS

UNDERSTANDING THE "BUZZWORDS"

Building owners, engineers, and other professionals associated with building design construction, operations, and management are overwhelmed by "buzzwords" related to networking and systems interoperability. Most, if not all, of the words, expressions, and definitions sound new, with mysterious and sometimes misleading meaning.

The users may be further confused by "professionals" — sales engineers, consultants, and others using the communications-related buzzwords without full understanding of their definitions. Communications engineers or systems integrators are seldom the first contact group with the users when it comes to selling an automation system. This is not to say that the sales engineers and consultants are trying to intentionally mislead the users. However, one must recognize that the whole discipline is fairly new to most in the building and building automation industry.

Facilities engineers and managers are involved in engineering, operating, maintaining, and managing facilities. Communications and systems interoperability issues are outside of their main focus and, to a degree, outside of their professional training. While the volume of information related to communications and systems integration is overwhelming (and sometimes conflicting and confusing), the pressure to understand the related issues has become a daily fact of life.

Building owners and engineers feel the need for interoperability of their systems. They also need professional advice concerning systems integration. System vendors can aid their customers in integration related to their own system. However, they have limited understanding of the "global" needs of the customer when it comes to integration of automation systems outside of their own systems (or category of systems, i.e., BAS) into a full-scale facilities network.

To make it more challenging, the entire communications industry is constantly evolving at speeds unprecedented in any other engineering discipline. This brings constant changes, leaving those of us who are trying to catch up with it behind. Furthermore, there is more than one solution to just about every communications or connectivity problem, providing for a wide variety of choices and options for implementation. Because of the constant evolution of the discipline and introduction of new products at unprecedented rates, yesterday's best solutions could quickly become today's "white elephants."

Most facilities automation systems and networks were developed over time. Therefore, the issue of interoperability among already installed systems, current projects, and future upgrades is the most relevant to the end users. **The challenge is to provide solutions that will lead to further development without obsolescence of the already installed systems.** Facilities engineers must either understand the related subjects or have a systems integrator or a networking professional on the design team. *Systems interoperability is a complex issue, and end users that think otherwise could make costly mistakes.*

All of us associated with facilities are experts in our own fields. Communications and interface experts are usually outside of the traditional facilities automation and environmental controls design and implementation teams. End users (applications and design engineers associated with building design, construction, and management) have no choice but to learn the meaning behind the communications buzzwords and to gain working knowledge of the disciplines associated with systems interoperability.

THE MOST COMMON MISREPRESENTATIONS OF INTEROPERABILITY

One of the most common "misrepresentations" of communications-related issues stems from the desire to simplify the subject matter at hand. We all have a tendency to do so, or we are led to do so by the situation or by the desire to satisfy our audiences. In an effort to simplify this complex subject ("...cut all that technical mumble-jumble and get right to the solution..."), too much is left to interpretation by people who are new to the subject. Since "free" interpretations are based on the specific background of the individual, understanding of not fully defined subjects (such as interoperability) will differ from individual to individual.

A prime example of simplification, when new systems or design solutions are presented to end users, is that the "new system will interface with other systems or with the existing ones via serial communications" — or, if the system or solutions are more sophisticated, "via an Ethernet protocol." By using the above commonly known expressions, the end users are supposed to be put at ease.

Why? Because, most end users are familiar with serial communications between their PCs and their peripherals, and they know they have worked. Some end users can relate to the Ethernet. Data communications people have used it in LAN networking for years. The Internet brought into our consciousness the TCP/IP protocol (transmission control protocol/internet protocol), or at least the IP part of it — that's how favorite Websites are accessed on the Internet. Now, if we can access a Website anywhere in the world from our PCs (thanks to the IP protocol), the system that claims interoperability via an Ethernet and TCP/IP should be good enough to network with all automation systems on the site, right?

The answer should be a reluctant "maybe," and a qualified "no." (Be warned; by giving such an answer, one is opening a veritable "Pandora's box" of endless questions and arguments with the audience. Be prepared for disenchantment of the audience with the details while desperately trying to explain the basics).

FUNCTION OF COMMUNICATION PROTOCOLS ON HUMAN COMMUNICATION

Even before us, the end users, communications experts have struggled with similar problems and definitions. Unable to unify or define a "standard protocol," they came up with the next best thing. They published a model for open systems interconnections (OSI) in 1977. The OSI model defines activities related to communication protocols in seven layers. The highest, the seventh layer, is interfaced to the application program, while the lowest, the 1st layer, is connected to the network media, such as twisted shielded pairs of wires or fiber optic cables. Most communication protocols in use today use either all or some of the seven layers of the OSI model (Figure 2.1).

If one relates the above statements concerning serial communications and/or Ethernet to the OSI model, one can see that defining serial (i.e., EIA or RS-232) or Ethernet communication for the systems, one has defined only the two lowest layers (first and second) of the OSI model. One would have done much better by defining the TCP/IP protocol, since TCP is on a transport (fourth) layer, and IP on a network (third) layer. As one can see, there are still several undefined layers,

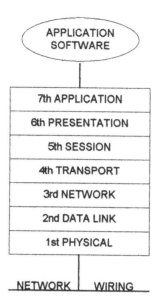

FIGURE 2.1   The OSI model.

even if the given protocol were to use a collapsed model (not all of the OSI layers used) for transfer of data from one computer or controller to another.

This is to say that definition of the lower layers of the OSI model provides only a communications path to the media, but does not guarantee interoperability of systems on the network.

To bring the subject closer to nontechnical readers, let us represent the communications protocol in a human environment. Between humans, the word **communicating** means that there are two or more people (nodes, in the computer world) with the desire to communicate with each other. **Protocol** means that to communicate they must observe some set of rules, behavior, and conduct.

If one is to examine the OSI model on humans (and we know that computers are designed after humans, by humans), one has to relate the layers of the OSI model to human communication. First, let's define the communication media (or **physical** layer of the OSI model). In a written communication (as I am communicating with you now), it is paper, just like the copper wire or fiber optics of the communications network. Second, I have decided to use the English language and the alphabet (as opposed to Chinese language and its alphabet). The alphabet in written communication can be equated to the electrical characteristics of a network, defined in the **data link** layer.

As can be seen, defining the first and second layers of communication alone does not warrant transfer of information to all readers (nodes on the network). For example, a Spanish-speaking person would not understand English communication, even though both the paper and the alphabet look familiar.

Consequently, deciding on a written communication, the use of the English language, and the alphabet, this book can reach only the English-speaking readers, unless it is translated into another language(s).

To communicate effectively with you, the English-speaking readers (clients or nodes on the network), I need to follow established conventions (or *protocols*) for technical books. The rules of conduct or protocol for technical books are defined by the English language, as well as by years of experience of authors, editors, publishers, and the readers striving for effective communication.

In a network communications they are defined by a "communication protocol or driver" used by all systems connected to the same network. Such protocol must then reside on every computer (node) connected to the network.

When I wish to communicate an idea to you over the agreed-upon medium, my brain has to format the outgoing messages (ideas) into English sentences which can be received and understood by you, the reader. This function is similar to the function of the **application** layer of the OSI model.

Next, we (the publisher and I) have to take care of other, protocol-related, issues common to technical publications. We have to make sure the ideas are presented to you in a format generally accepted by readers of technical publications (text, graphs, tables, sketches, etc.). This is similar to the **presentation** layer of the OSI model.

The publisher and I also have to structure the ideas into sentences, clear and concise paragraphs, pages, and chapters. These structures are similar to the function of the **session** layer of the OSI model.

The publisher is also responsible for editing the book so the information is printed in an error-free format. This function is similar to the function of the **transport** layer of the OSI model.

The ideas formatted into error-free sentences, paragraphs, and chapters, are double-checked again prior to setting them into the English alphabet, and checked again prior to printing the book pages. This function is similar to the **network** layer of the OSI model.

The above "protocol" is utilized in the printing of English language technical books all over the world.

Figure 2.2 provides graphical presentation of two nodes communicating on the same network, utilizing the same protocol. Protocol functions, residing in the software and hardware, are shown on the left.

Now, this situation would change if there were to be "foreign" nodes with different protocols connected to the same network with an intention to communicate with other nodes on the network. For example, a Chinese reader (let's equate it to a foreign node on the network) cannot read or

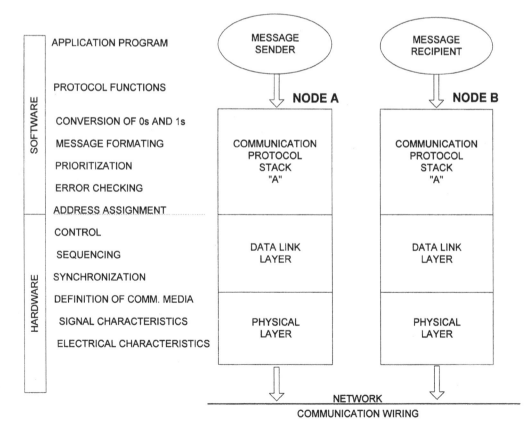

FIGURE 2.2   Example of network nodes using the same protocol.

understand the printed English text in this book (called protocol "A" in Figure 2.2). Despite his/her familiarity with the communication medium, the paper, he/she would not be able to read nor comprehend this text.

In publishing, there are several options to reach the "foreign" readership:

1. Teach all engineers of the world the same language — English, in this example. This would give them an ability to communicate with all engineers on this planet (a common communication protocol approach).

However, this would leave out all nonengineering readership, like managers, architects, and other professionals associated with the building industry. This can be equated to implementing BACnet protocol for BAS without interfaces to other facilities systems. By doing so, all other facilities systems, such as, for example, maintenance management systems, industrial control, and automation systems for power plant automation, are excluded from communicating on the same network. Graphical presentation of this approach is in Figure 2.2. This approach would require inclusion of a common protocol "A" to all nodes, regardless of systems of different categories connected to the network.

2. Translate the text of the book to foreign (i.e., Chinese) languages, a common approach adopted by publishers.

This approach is also adopted by most systems integrators to interface systems of different types to each other in absence of a "standard" facilities protocol.

Adopting this approach of integration of two (or more) systems in absence of a common communications protocol means adding a protocol translator(s) to a node on the network, or adding a network device (a server, a PC, etc.), to provide the protocol translation for the foreign network devices. Such a network device can contain more than one protocol, thus providing interoperability for greater variety of "noncompetitive" systems. The example in Figure 2.3 shows a client-server architecture. Noncompetitive systems (i.e., BAS and power plant controls and automation systems) communicate via a server. The server (i.e., Intellution's FIX) can also provide uniform interface and data presentation to the clients connected to the same network. This is called a SCADA client-server type of communication.

The above, "human" interpretation of the OSI model, is to aid in visualization of the importance of individual layers of protocol for error-free data communications and for interoperability between systems connected to the same network.

Time and systems development will undoubtedly change the landscape of networking and systems interfaces in the future. However, responsibility for systems interoperability, selection, and engineering of the most appropriate protocols and drivers will remain with the teams associated with building systems design, implementation, and management.

COMMUNICATION PROTOCOLS

To communicate from one computer to another (Figure 2.4), one has to establish **connectivity** and **interoperability** over a network. To transmit, receive, interpret, and acknowledge a message over a network, it has to be in a format that is understood by computers or controllers on the network.

Rules governing procedures related to data transmission over a network are specified in a form of a **communications protocol** or **driver**.

There are three groups of protocols/drivers: **Proprietary**, **Open**, and **Standard**.

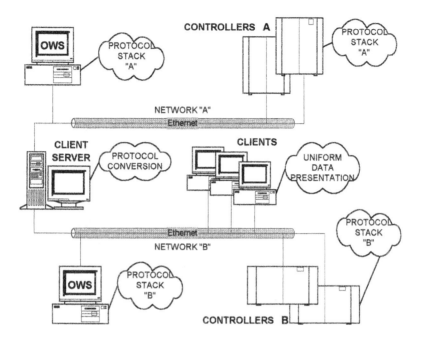

FIGURE 2.3 Client-server approach to communication.

PROPRIETARY PROTOCOLS

Proprietary protocols are developed by systems or computer manufacturer(s) to communicate to their own (proprietary) hardware and software over a recommended network. They are specifically designed and thoroughly tested for vendor-specific system(s) and for specified networks. Since proprietary protocols are not published, other systems or computers cannot coexist or communicate with the vendor-specific system or computers on the same network. Understanding the above is important for development of facilities automation systems.

On one hand, proprietary nature of systems assures systems reliability and integrity; on the other hand, it precludes the end user from utilizing other vendor's off-the-shelf controllers or systems on the same network. Proprietary systems competitively priced, well-supported, and on a par with developments in the automation industry, have gained owners' loyalty over time. While many end users feel loyalty to their proprietary systems and vendors, an equal number of customers is disenchanted with proprietary systems and vendor support and feel locked into XYZ's proprietary automation systems.

Proprietary protocols have evolved over time with the computer and automation industry. They have their place in the market, even with the growing desire for systems interoperability. Every facility has mechanical equipment and systems that require automation with a high degree of systems security, integrity, and with very little or no interaction with other systems on common facilities networks.

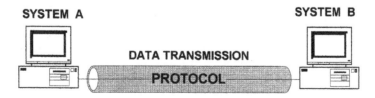

FIGURE 2.4 Communication from one computer to another.

Utilizing proprietary protocols is comforting for facilities managers and engineers for one simple reason: the responsibility for the automation systems operation and troubleshooting, as well as for systems and communication integrity and reliability, is with one vendor.

OPEN PROTOCOLS

Many open protocols started out as vendor proprietary protocols. They became public domain when the vendors invited other system developers to write interfaces and share data on their network. Opening up proprietary protocols means disclosing procedures, structures, and codes to systems integrators designing other systems to communicate on the same network.

Open protocols became a fact of life in the computer industry, in office automation, in industrial controls, and in building automation applications. How popular or widely accepted an open protocol or a communications driver becomes depends on its quality, features, and provided services. Most important, how are they accepted by systems engineers and end users? Their acceptance follows a path of "natural selection;" they represent the best competition can produce.

The best of them became so-called **de facto** or **industry standard protocols** or **drivers** (not standard protocols), widely utilized by systems designers and system integrators. The advantage of industry standard protocols or drivers is in their performance, successfully tested in many applications and refined by many improvements over time. The disadvantage, at least for the end user, is that most systems vendors and many systems integrators have developed and market one or more such interface driver or protocol. The previously narrow selection of one meal on the menu became one meal selection from a smorgasbord of the automation industry. Careful selection from the menu may provide a very satisfying meal or an upset stomach, if the customer lacks familiarity with the offerings.

Most DDC systems on the market today have open protocols or use "de facto standard" communications drivers. Many major vendors offer interoperability, with tens of systems related to building HVAC systems and an opportunity for interface protocols and drivers to be written by everyone interested. They have shifted the burden for systems interoperability (and many times for protocol testing) to the end users or systems integrators. End users should be aware of problems associated with systems integration based on unproven technology and untested protocols. A laboratory bench test is not a substitute for testing out a communications protocol in a real environment.

DDC system vendors provide a list of so-called industry standard protocols and drivers, which allow third-party controllers and systems to interface to their system. Beware — such interfaces are usually "one-way" interfaces. Many such interfaces may have been developed and tested for a specific job. End users or integrators should require the vendor of such protocols or drivers to provide description of the protocol or driver features and services, including results of performance testing. They also should visit the site where such an interface was implemented and match the protocol performance with their own site application requirements.

Proprietary, open, and industry standard protocols represent a "grass roots" movement, a development initiated by the computer, controls, and automation industry. The latter two also represent the desire of the end users for systems integration.

STANDARD PROTOCOLS

Standard protocols represent the effort of the entire industry (or selected groups) to bring order to the ever-evolving computer, communications, and automation industry. Each of the related industries (such as manufacturing, building automation, industrial controls, etc.) made an attempt to develop standards that would closely meet their specific needs and requirements. Most, if not all, such protocols are based on a 1977 model for open systems interconnection or OSI published by the International Standards Organization (ISO). OSI is a model (not a protocol) adopted by the computer industry, which categorizes the activities essential for data transfer from one computer to another over the network.

UNDERSTANDING THE OSI MODEL

The OSI model defines network activities in seven layers. These range from the highest layer (seventh) of software activities (such as issuing a command or displaying data) to the lowest layer of transmission of the signal over the physical network.

Application Software

7th layer	Application
6th layer	Presentation
5th layer	Session
4th layer	Transport
3rd layer	Network
2nd layer	Data link
1st layer	Physical

Network

In the model, each layer has an associated address, and processes data delivered from the layer below or above, or from the same layer of another computer or node on the network. The upper (software) layers are designed to deliver exported data to and from the application software; the lower layers are designed to interface data — in a form of an electrical signal — to the physical network. Lower layers of some protocols utilize so-called "silicon integration" for network connection, which could also be implemented on a single chip.

LAYER 1 — PHYSICAL

The physical layer specifies the communication path and physical media of the network. Among the features are definitions of:

- The actual transmission media, such as twisted pair wire, coaxial cable, fiber optics cable, etc.
- The electrical signal levels and drive capacity
- The characteristics of the media, such as transmission wire length and the speed, and amount of data it can handle without a need for signal regeneration

The parameters of the physical layer are standardized by the *Electronics Industries Association (EIA), Recommended Standards (RS)*, and the *Institute of Electrical and Electronics Engineers (IEEE) standards* (Figure 2.5).

The most widely used standards in the automation industry are the IEEE 802 standards, and RS-232, and RS-485 standards for serial communications.

LAYER 2 — DATA LINK

The data link layer controls access to the physical layer and network media. The primary functions of the data link layers are

- Control, sequencing, and synchronization of transmitted data (sending and receiving)
- Low-level error detection and error recovery
- Timing and other functions associated with control of the physical media

IEEE 802.1 HIGH LAYER INTERFACE	HIGHER PROTOCOL LAYERS
IEEE 802.2 LOGICAL LINK CONTROL	DATA LINK LAYER
IEEE 802.3 CSMA/CD MEDIA ACCESS IEEE 802.4 TOKEN BUS MEDIA ACCESS IEEE 802.5 TOKEN RING MEDIA ACCESS OTHER IEEE STANDARDS	PHYSICAL LAYER

FIGURE 2.5   IEEE standards associated with physical and data link layers.

IEEE standards divide the data link layer into:

High level data link control (HDLC)
Logical link control (LLC)
Medium access control (MAC)

While the higher layers, including HDLC and LLC, are typically implemented in the software, MAC (and the physical layer) are usually implemented in the hardware (most typically as boards added to the PC or as communication modules added to controllers). The hardware usually provides bit handling such as encoding, error detection and recovery, address detection and recognition, and functions associated with electrical signal levels and drive capability.

Layer 2 also determines which methods of communications control (master–slave, peer-to-peer, hybrid) are used on the network, provides error handling (bit checking, cyclic redundancy checking — CRC), and provides control of communication (data) flow.

LAYER 3 — NETWORK

The network layer (Figure 2.6) provides interfaces between the transport layer above and the data link layer below in the OSI model. The network layer is responsible for establishing and termination of connections between the originator and recipient of information over the network. The network layer assigns unique addresses (numerical codes) to each computer (node) on the network. The addresses

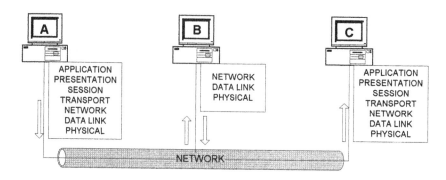

FIGURE 2.6   Network layer, node-to-node communication.

also identify the beginning and end of the data transmission packets. The network layer is responsible for delivering the information directly to its destination or finding alternate routes, in case of problems with intermediate computers residing on the network, if guaranteed delivery of a packet is required.

LAYER 4 — TRANSPORT

The transport layer (Figure 2.7) maintains reliability on the network and enhances data integrity. The layer provides error recovery (for example, by requesting retransmission of incomplete or unreliable data). The transport layer is concerned with getting the data to an intended address. It enhances the reliability of the network layer by focusing on getting the data reliably to the other computer's or node's transport layer, or saving it in case of communication failure.

There are four (4) so-called integrity classes guiding error checking and error recovery: the lowest (Class 0) provides very little error checking, while the highest (Class 4) provides extensive error checking and recovery.

There are several ISO standards associated with this layer, as well as other standards, such as X.25 Wide Area Protocol and the Transmission Control Protocol (TCP).

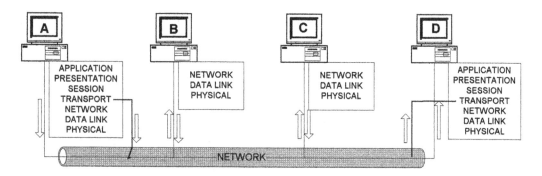

FIGURE 2.7   Transport layer, node-to-node communication.

LAYER 5 — SESSION

The session layer establishes (sets up and terminates) and manages sessions between application programs over the network by synchronizing the data exchange. The session function is important for PC network communications, managing access to application software resident on different PCs or on a network server. The session layer also provides prioritization (high, normal) and additional software error recovery of data.

LAYER 6 — PRESENTATION

The presentation layer's main task is to convert data received over the network (from different protocols) to a format that can be easily utilized by the application program. It provides data formatting, translation, and syntax conversion of alphanumerical (character) as well as graphical information. The layer is responsible for receiving, unpacking, decoding, and translating data into formats and codes (names, objects, entities, graphical symbols, etc.) understood by the application software.

LAYER 7 — APPLICATION

The application layer provides a variety of services utilized by the application program. The services, whether application or user specific, allow for data transfer and management, file access, time stamping, remote device identification, and other high-level service functions. The application layer allows application programs to use the lower layers (1 to 6) in a generic way, regardless of the protocol stack below.

ISO 8650 COMMON APPLICATION SERVICE ELEMENTS ISO 9041 VIRTUAL TERMINAL	APPLICATION
ISO 8823 CONNECTION-ORIENTED PRESENTATION PROTOCOL SPECIFICATION ISO 8825 BASIC ENCODING RULES FOR ASN.1	PRESENTATION
ISO 8327 CONNECTION-ORIENTED SESSION PROTOCOL SPECIFICATION	SESSION
ISO 8073 CONNECTION-ORIENTED TRANSPORT PROTOCOL SPECIFICATION ISO 8602 PROTOCOL FOR PROVIDING CONNECTIONLESS TRANSPORT SERVICE	TRANSPORT
ISO 8373 PROTOCOL FOR PROVIDING CONNECTIONLESS NETWORK SERVICE TRANSMISSION CONTROL PROTOCOL	NETWORK
IEEE 802.2 LOGICAL LINK CONTROL MEDIUM ACCESS CONTROL HIGH LEVEL DATA LINK CONTROL	DATA LINK
IEEE 802.3 CSMA/CD IEEE 802.4 TOKEN PASSING BUS IEEE 802.5 TOKEN PASSING RING RS 232, RS 485 SERIAL COMMUNICATIONS	PHYSICAL

FIGURE 2.8 Standards associated with layers of the OSI model.

Network layers communicate with each other vertically (layer n to layer $n+1$ or layer $n-1$) within the same computer, or horizontally with the same layers of another computer via a **network**.

There are numerous protocols developed by the computer as well as the controls industry, utilizing the ideas and basic structures of the OSI model. The seven layers of the OSI model are the base for so called **open systems** or **standard communications protocols**. Protocols of computer systems, including DDC systems, adhere to the OSI model or its modifications, called "collapsed models," which utilize some of the layers, or combines two or more OSI layers into one layer of the collapsed model.

Figure 2.8 is an illustration of standards associated with individual layers of the OSI model.

PHYSICAL MEDIA AND NETWORK TOPOLOGY

Software-generated messages from one computer to another have to be digitized (converted to 0s and 1s) and transmitted over the network in the form of electrical signals. Communication signals travel on conductors (coaxial cables, fiber optic cables, twisted pair wires, etc.), or can be transmitted over other media such as radio frequency (RF) and infrared signals (IR), or can be modulated on AC power lines (called power line carrier). The media of transmission are called **communications** (or **physical**) **media**. The most common communications media utilized by facilities are cables (coaxial, fiber optic, and twisted pair cables).

The word **network** is often associated with the physical media (communications cables) and its topology in mind of the general public.

Physical Media — Communications Cables

Twisted Pair Cables

These are used by telecommunications (multipair telephone cables) as well as by controls and automation vendors (multi- or single pairs). Multipair telecommunications cables (1 mm copper) are used for up to 1.5 Mbps transmission speed for digital communications.

Twisted pairs (20 to 26 AWG) used in facilities automation are shielded to prevent electrical interference from installed electrical equipment and other sources of induced noise.

The greatest advantage of **twisted pair cables** in local area network (LAN) communications is their availability in the existing facilities and their low cost associated with a node connection. Twisted shielded cables are popular in DDC communications (controller to controller, or controller to sensor), due to their low cost and ease of installation. The disadvantages are in their susceptibility to electromagnetic interference and low transmission speeds.

Coaxial (coax) Cables

These are used by telecommunications (LANs) as well as in networking automation systems (i.e., for controller-to-controller communications). Coax cables are constructed of a copper core, insulating material, outer conductor (braided or meshed), and a protective outer insulation. Coax cables are used for baseband as well as broadband transmissions, due to their effective bandwidth ranging from 100s to 1000s of MHz. Digital transmission speeds are up to 50 Mbps on single channel baseband transmissions. Common cable impedances are 50 Ohm (Ethernet), 70 Ohm, and 93 Ohm (ARCNET). Coax cables are more expensive and more difficult to install and terminate than twisted pair cables.

Fiber Optic Cables

Optical fiber has earned increased popularity due to improved speed of digital transmission and increased lengths without a need to repeat the signal. Fiber optic networks operate in a range of 10s of 100s of Mbps, with some systems up to 1.7 Gbps, all with much better error rates than those in copper cables. This is due to immunity of fiber optics to electrical noise and interference. Fiber optic cables are still more expensive than copper and much more difficult and expensive to terminate and splice. In the majority of telecommunications installations (LAN backbones), fiber optic cables are the preferred choice for network operators.

Installation and operating costs, transmission speed, maximum cable lengths, network security, and reliability are the determining factors for selection of the physical media for facilities networks and automation systems communications.

NETWORK TOPOLOGIES

Network layout or **topology** is described by the geometrical configuration of computers (or nodes) connected to the network. The base topologies are bus, star, and ring topologies.

Bus Topology

Bus is a "continuous cable" to which network devices or nodes are connected directly (Figure 2.9). Data in a bus topology are passed from one node to all other nodes without any intermediate device. Bus topology is easy to implement and expand. Many Ethernet systems utilize bus topology. Since

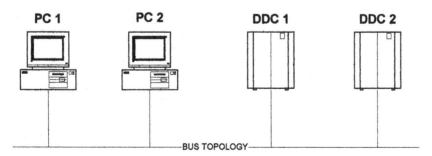

FIGURE 2.9 Bus topology.

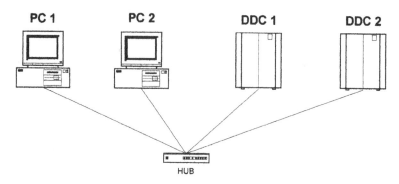

FIGURE 2.10   Star topology.

the bus is a parallel communication system, failure of a single node will not affect the other nodes connected to the network.

Star Topology

Computers in star configuration communicate with each other via a central device — a hub (Figure 2.10). This was a common topology of centralized computing; terminals or computers were connected to the central computer in a star configuration. Telephone private branch exchanges (PBXs) utilize star configuration with a central switch connecting all nodes of the network. Reliability of a star topology is reduced by a "bottleneck" — the hub, a single point through which connected nodes communicate with each other.

Ring Topology

Nodes on a ring network (Figure 2.11) are connected in series. Each node has another one on each side forming a closed loop. Data sent from one computer to another goes through all computers connected to the ring until it comes back to the sender. Single failure of one device on the ring may not affect the other nodes on the network if the nodes can reroute communications bidirectionally on the broken ring.

The above basic topologies have their variations — Bus-Branching Tree (Figure 2.12), Star-Hierarchical Levels (Figure 2.13), and Star-Wired Ring (Figure 2.14).

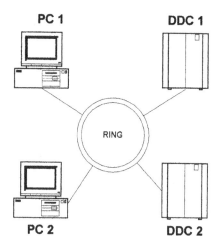

FIGURE 2.11   Ring topology.

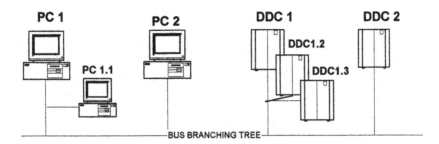

FIGURE 2.12 Bus-branching tree.

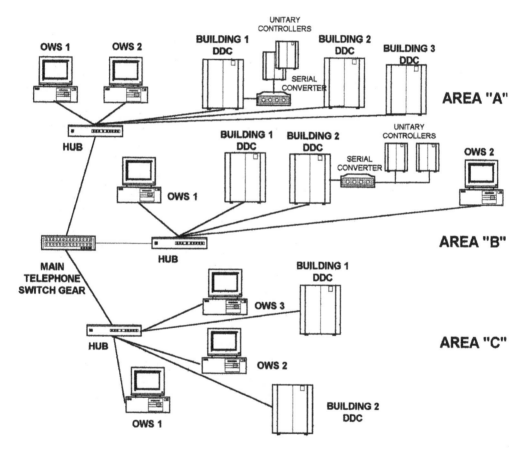

FIGURE 2.13 Hierarchical star topology.

DIGITAL ENCODING

Every computer (PC or controller) transmitting data to others over the network or transmitting data to their respective I/O devices (i.e., printers) converts characters or messages into a digital format, encodes them, and puts them on the physical media as electrical signals. Digital conversion is common not only for alphanumerical characters, but also for voice and graphics transmission.

Encoding of digital data assures improved, reliable data communications on a network. Common encoding techniques improve noise immunity and error detection.

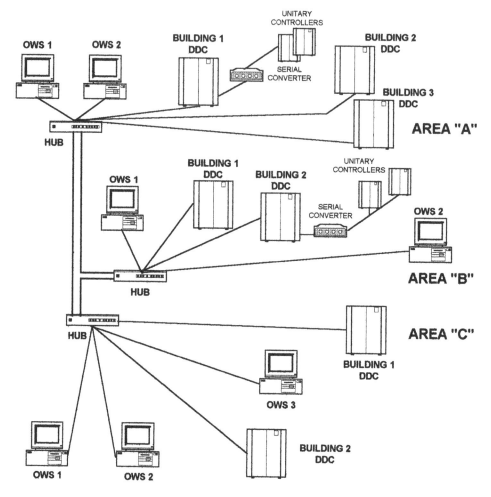

FIGURE 2.14 Star-wired ring.

The following are the most common encoding wave-forms: pulse return to zero (RZ) or nonreturn to zero (NRZ); unipolar (pulse returns to zero), polar or bipolar (pulse polarity changes from + to – value); and Manchester and Miller digital signal encoding, pulse width per bit time (or half bit time), are used more with advancement of common communication protocols (Figures 2.15 and 2.16).

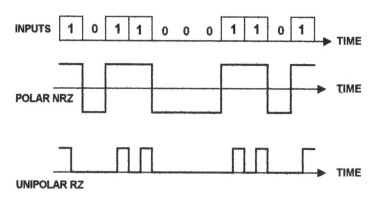

FIGURE 2.15 NRZ digital signal encoding.

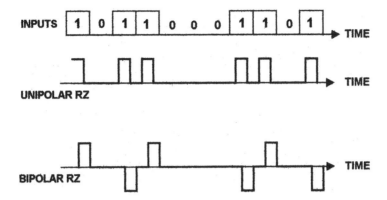

FIGURE 2.16 Unipolar and Bipolar RZ digital signal encoding.

SIGNAL MODULATION

Binary signals (0s and 1s) can be unmodulated carrier; amplitude modulation (amplitude is varied by multiplication of 0s and 1s); frequency modulation (different frequency for 0s and 1s); and phase modulation (the phase is shifting from 0 up to 180 degrees) (Figures 2.17 to 2.20).

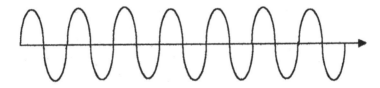

FIGURE 2.17 Unmodulated carrier.

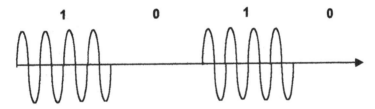

FIGURE 2.18 Amplitude shift keying (ASK).

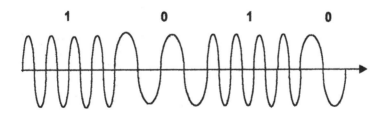

FIGURE 2.19 Frequency shift keying (FSK).

BITS, BYTES AND BAUD RATES.

Bits (bps) express data transmission rates per second for data communications.
Bytes (B) are groupings of bits (most commonly 8 bits) into one group.

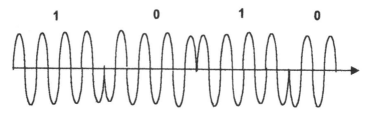

FIGURE 2.20   Phase shift keying (PSK).

Baud rate (baud) is a number of signal changes per second (amplitude and phase changes) transmitted over a transmission line (i.e., NRZ encoding uses one baud per bit, Manchester uses two baud per bit of digital signal).

PARALLEL AND SERIAL TRANSMISSION

In **parallel transmission** every bit is transmitted on its dedicated wire (transmission path) simultaneously. This results in high throughput. Parallel transmission is common between computers and their peripherals located in their close proximity.

For **serial transmission**, bits are assembled into bytes and transmitted in a single transmission path. Serial transmission is most suitable for communications to remote locations, but is somewhat slower than parallel transmissions.

ASYNCHRONOUS AND SYNCHRONOUS DATA TRANSMISSION

In **asynchronous transmission** mode, the computer initiating transmission sends out a start bit, followed by a single byte and one or more stop bits, if using ANSI standards. The bit transmission occurs at set time intervals provided by an internal clock. The receiving computer recognizes the start bit and reads the incoming data (bit stream) at a fixed time interval, resynchronizing its internal timing with each bit and at each start bit. Transmission errors may be detected by so-called parity bits at the end of each byte (odd parity means that the total number of "1" bits is an odd number; even parity means that the total number of "1" bits is an even number).

In **synchronous transmission** mode, the clock rate is part of the bit stream sent by the transmitting computer. The receiving computer adjusts its internal clock rate according to the encoded rate in the bit stream. Bits are grouped into logical groups (frames or blocks) with embedded bits for synchronization. Error detection is usually provided by the use of several bits at the end of blocks of data (i.e., block check character or cyclic redundancy code).

SIMPLEX, HALF-DUPLEX, AND FULL-DUPLEX MODES OF DATA EXCHANGE

Simplex data exchange utilizes a single path (pair of wires) in one-way data transfers.

Half-duplex data exchange is bidirectional, but allows transmission in one direction only at a time.

Full-duplex data exchange provides for bidirectional, simultaneous data exchange.

Data exchange in half-duplex and simplex modes can be via 2-, 3-, or 4-wire transmission paths. Data exchanges using full-duplex mode require 3- or 4-wire transmission paths.

TRANSMISSION TECHNIQUES

Two major classifications of LANs are used: **baseband** and **broadband**. They are identified by their speed of transmission, access methods, and topology.

BASEBAND TRANSMISSION

Baseband transmission rates are around 10 to 100 Mbps. Baseband communication is characterized by one signal being transmitted at any given time over the physical media. The signal is represented by changes in DC voltage (NRZ signaling). Baseband networks utilize time-sharing or multiplexing of resources for accessing the media (cable) by connected computers. Controlling the data transmission and assuring equal access of network devices (computers) to the network is by several **access methods** developed for LANs.

Topology	Access Method
Bus network — each network device (computer) is connected to each other by a network cable (pair of wires, coax or fiber optics)	**CSMA/CD** (carrier sense multiple access with collision detection) is used by *Ethernet* A CSMA/CD protocol in a transmitting network device is trying to detect (by listening for another signal on the network) a carrier, indicating that the network is used, before allowing transmission. It listens to the carrier during the entire transmission to detect a collision. If collision occurs, the network device stops transmitting for a random time before retrying.
	Token bus — a token bus access method establishes a sequence of network devices passing the "token" from one network device to another (in a round-robin sequence). If the network device does not want to transmit, it simply passes the token to the next node, until the token is returned to the network device that originated the token. Each device is time-limited for the number of transmissions it can do before it has to transfer the token.
Ring network — each network device is connected to the next one (the last network device is connected to the first one)	**Token ring** — the token is circulating from one network device to another one on a ring network. Each network device receives the token and retransmits it with a small (usually 1 bit) delay. The transmitting network device can change the token to a connector by appending any transmitted data it wishes to send to the token. Upon receiving the token back (i.e., the packet has circulated around the entire ring), it restores it to a token which then can be used by the next network device.
Star network — has all network devices connected to the active communications hub which controls the access right and provides packet switching (an example is a PBX — private branch exchange). Star topology can also utilize CSMA/CD and token ring access methods	**Star** — access of individual network devices is provided by individual line cards in the hub. The hub provides a path connecting the transmitting network device to the designated receiving network device and maintains it for the duration of the transmission.

BROADBAND TRANSMISSION

In **broadband** transmission, the transmission frequencies are divided into different channels. The speed of transmission depends on the used physical media (cable). Fiber optics cables can transmit at a speed of several Gbps to distances of several miles without signal repeaters. Broadband transmission is typical in broadcasting and metropolitan area network communications. Broadband transmission allows the use of the same cable for analog signals (such as TV signals) with digital data transmission. In fact, its biggest advantage is its ability to allow many different systems to communicate on the same network, such as network terminals to a central computer, PC to a PC or server, PC to DDC controllers, closed circuit TV, telephone, and other systems. Broadband channels for standard analog (color television) transmission require a 6-MHz bandwidth; for digital transmissions it is 300 kHz to 20 MHz, depending on signal encoding and required transmission speed.

The broadband topology and access methods are similar to the ones in baseband communications.

NETWORK SECURITY AND MANAGEMENT

Security of networks and of the connected network devices (computers) is becoming an increasingly important issue. In data processing, breach of security (an unauthorized access) can result in illegal data manipulation or file corruption. In real-time systems such as BAS and power plant control systems, an unauthorized command (for example, to start a large chiller) may result in damage to the equipment and injury to people working on the equipment.

Considerations for network management should be part of the network development process. As facilities networks get more complex, a higher level of expertise is needed for their management. Since network management was traditionally provided by telecommunications specialists, facilities attempting to develop their own networks must plan adequate allocation of manpower and sufficient funding for network management.

THE BENEFITS OF UNDERSTANDING INTEROPERABILITY

EDUCATING OURSELVES

End users benefit from understanding interoperability and networking in many ways. The greatest benefit is from educating oneself. Educated end users can make better decisions relating to long-range development of facilities automation and their integration. Understanding interoperability issues could reduce the cost for networking directly by specifying the most appropriate hardware and software and by eliminating unnecessary expenses for changes and redesign of systems and networks.

The benefits educated end users can bring to the organization are listed below. Educated end users are

- More qualified to provide or evaluate proposed solutions which will lead to further development of their facilities automation systems without obsolescence of already installed systems and networks.
- More qualified to develop or evaluate master plans related to integration and development of their facilities automation systems and networks, to provide realistic financial proposals and schedules to the upper management, and to evaluate cost and benefits of proposed networks and systems integration.
- Able to relate to vendor presentations, which could lead to realistic specifications and cost estimates for establishing interoperability among systems. Such understanding of interoperability is important, even in cases when there are qualified controls engineers on the job, since their interests and responsibilities are directly related to the job at hand, but not to the overall facilities network.
- Better qualified to safeguard the integrity of systems and networks as specified in the facilities master plan. The qualified end user can review proposals of individual projects

over extended periods of time, and assure that the proposed systems and networks are interoperable with the systems and networks already installed or being planned for installation. Some of the items to review are

- Controllers, PCs, servers, etc.
- Operating systems
- Interface hardware and software
- Communication media and network devices
- Protocols and drivers
- Applications software
- Applications programming, including programming related to interfaces
- Hardware and software installation
- System validation, end-to-end testing, documentation
- System warranties, documentation, and training

OPTIMUM UTILIZATION OF EXISTING SYSTEMS AND NETWORKS

Facilities, especially existing facilities, have in place systems and networks which could be reused for new installations. Cost of field installation is one of the highest (if not the highest) items for any systems implementation. There are definite benefits in reusing existing wiring or existing spare data grade telephone lines for communications. Many large facilities have communications equipment or networks that also can be used for real-time data communication.

SELECTION OF THE OPTIMUM NETWORKS FOR EACH LEVEL OF INTEGRATION

Not every controller has to be on a high speed (and high cost) network. For example, if there is an Ethernet in the facility, select DDC vendors, which can utilize Ethernet for building-to-building communication. It could save money over selecting a vendor, for example, with an ARCNET communication. If, for whatever reasons, an ARCNET is selected, one should add to the cost of the project the cost for a dedicated network or for ARCNET-to-Ethernet protocol conversion.

Communications between individual application-specific controllers can be accomplished by serial communications over existing data-grade telephone lines or dedicated wiring, rather than by installing a high speed network.

DEFINITION OF THE MOST APPROPRIATE INTERFACES

Facilities with automation systems installed over several years have a lot of money invested in their systems. Some of the systems serve their purpose even without interoperability. When deciding on which systems to integrate and how, the end users should evaluate their existing installations and decide on the most economical course of action.

The following is a list of possible options.

- **Migration of Systems:** System vendors provide migration paths for their systems. There is a good chance that an older generation automation system can be migrated into a newer generation system (from the same vendor), which is easier to integrate into the facilities network. It will probably also be less expensive than trying to interface an older generation system into a new network.
- **Utilization of Interfaces within the Same Category:** For example, your DDC system may have an interface to another DDC system; or you may select vendors that use the same interfaces, such as (for example) the BACnet, CAB, or LonTalk®.
- **Development or Utilization of "De Facto Standard" Interfaces between Systems of Different Categories:** Utilization of a client–server arrangement, with servers residing

on the facilities network, interfacing to a number of DDC, utilities metering, power plant, or maintenance management systems, provides a cost-effective solution to networking. First, system integrators offer interfaces to a variety of systems using custom written or "de facto industry standard" protocols or communication drivers, such as (for example) Modbus interfaces. There are also other interfaces used by the data processing and industrial controls industry that can be utilized for mapping over points from individual systems to the server. Second, most of the clients in this scenario are already connected to the network; therefore they already have access to the network. Third, with advancement of the Internet, vendors offer interfaces via their own web servers. One should be using all of the available options to achieve systems integration and distribution of data to the customers and staff of the facilities department. However, options have to be utilized with care to prevent data corruption and unauthorized access to the network, data, and systems.

- **HVAC, DDC, and Mechanical System Retrofit:** Control systems, just like the mechanical systems they control, have a limited life cycle. The times are gone when a mechanical governor controlled a steam engine for many decades. Control systems and their interfaces are antiquated at a much faster rate. Interfacing older DDC systems to a more modern communication system or to other DDC systems of the present generation can be costly or impossible. This is usually the case at the end of the control system's useful life cycle. Such systems should be retrofitted by newer systems, which assure interoperability over the facilities network. System retrofits can be coordinated with retrofit of the mechanical systems they control, which is usually easier to justify, due to resulting energy or maintenance cost savings. Some control systems should be retrofitted as part of the building or space renovation.

STANDARDIZATION

One of the benefits resulting from understanding interoperability is being able to standardize systems and networks for the facility.

Following are a few categories to focus when considering standardization.

- **Controls and Automation Systems:** As we know, they are expensive. They require maintenance by a trained work force. They also require spare parts, not to mention ongoing support. All of this costs money. In fact, the O&M cost of any system far exceeds its initial cost for implementation. This cost can be greatly reduced and the system reliability greatly improved by standardizing on a limited number of carefully preelected systems. The savings alone (not to mention the reliability of operation of systems maintained by a well-trained work force) could be much greater than the perceived savings resulting from the bidding process — apart from preventing an O&M nightmare and a mess created by selecting the lowest cost-control system for every project in a bidding process.
- **Applications Software:** It is very tempting for a systems engineer to pick up the latest and greatest software for every job. But remember, systems have to be supported. Ask yourself how many pieces of software can *your* operators learn; how many can *your* mechanics and technicians maintain; can *you* provide adequate and continuous support for all the applications software that you thought it would be nice to have for the last decade or so? We are talking about full understanding and a working knowledge of the software (so the operators would be able to troubleshoot and modify systems definitions and parameters), not just familiarity with it. Again, standardization saves money by having a well-trained support group fully familiar with the application programs who are being able to troubleshoot system problems before they become known to the clients.

- **Server:** It is unlikely one can implement a large-scale facilities integration without having one or more servers on the facilities network. A server–client arrangement is a convenient way to provide information to the clients and customers. To provide uniform information presentation, we have to standardize on network servers. Along with that, we have to standardize on the client workstations (PCs) and their operating systems. Standardization provides user convenience, network integrity, and reduces the cost for network and system management, maintenance, and upgrades.
- **Communications:** It is very costly to set up and operate incompatible networks with inadequate and incompatible protocols and communications media. To provide adequate network security and management of discontinuous networks is a great concern to all associated with network development, maintenance, and management. Standardization on communications networks and protocols translates to more reliable networks, which are also easier to support and maintain. Analogy can be drawn to DDC systems operated and maintained by a group of people who are confident, understand the peculiarities of the system, and are able to maintain it with minimum downtime.
- **Protocols and Drivers:** The same considerations for standardization of operating systems and application programs apply also to standardization on protocols and drivers. At this state of technology, protocols and drivers are more temperamental than controls applications programs. They need to mature, just as DDC applications programs have matured over the last 2 decades. Standardization means less headache, a shorter learning curve, and less time for design, implementation, and support. Validation and testing of protocols and drivers by engineers who have full understanding of their behavior is a great money- and time-saver during systems startup.
- **Operator/Client Interfaces:** Even the best system is going to fail, if the operators or clients are not going to use them as intended. It is important to provide interfaces the operators and clients can relate to. For example, if the users are accustomed to Microsoft® Windows, try to provide data presentations on Windows. Customers will appreciate the familiar interface. The same is true for report generation and report formats. If the clients on the network are accustomed to spreadsheets, import the report data to spreadsheets. They will be using the data more often and will be able to customize and modify the reports, creating custom reports, charts, graphs, etc., not to mention that such customers will become devoted supporters of the system and of future network developments.

Since systems integration is an ongoing development effort for facilities in development, it is important to have educated users, who can foresee the directions their facilities networks will take and can support integration over long periods of time. In the long run, such continuous development will save money for the end user.

3 DDC Open Systems: An Overview*

John J. McGowan

CONTENTS

INTRODUCTION

The direct digital control (DDC) market is global and is filled with multinational customers and service providers. This chapter gives a broad perspective on issues at hand.

There are common needs for communication and control on the part of building owners worldwide. As these individuals seek to address these needs, the first thing they encounter is a baffling array of new concepts and technologies in the area of DDC communication and networking. As with every paper that I have published on this topic in the last 15 years, my goal is to address this issue by clarifying both the present and key milestones in the past that have led us to this point in time. This foundation will provide a useful framework within which to view what's ahead.

System owners began to voice concerns in the early 1980s regarding some complexities in the long-term management and expansion of DDC systems. In many respects, the DDC industry was in its infancy at that time and the rapid pace of product development combined with the breadth of products in the market presented issues for owners. The ultimate impact of these issues was a significant awareness of the role that communication played in the long-term success of DDC systems, and with that the importance of protocols and networking. These terms, along with countless communication related product and technology buzzwords, have become critical to system selection in the past 4 years.

Why 4 years? Because BACnet, an American Society of Heating, Refrigeration, and Air Conditioning Engineers (ASHRAE) standard, was published in 1995. Prior to that time, for more than a decade the issue was a source of varying levels of interest, but with the publishing of that

---

* Reprinted with permission from *Strategic Planning for Energy and the Environment*, a publication of the Association of Energy Engineers, Atlanta, GA, 1998.

standard, industry took notice. The standard, simply by virtue of its existence, should be able to solve a host of problems. As is often the case, some even believed it would solve problems it was never intended to address. To make matters even more complex, rather than having the numerous communication protocols in the industry resolve into one ideal option, the industry is facing several "ideal" options. A term that has evolved recently to refer to systems using one of the many proprietary protocols available in the market until now is a **legacy** system. So, if you have been confused about BACnet, open systems, LonWorks®, etc. or are wondering where the industry goes from here, read on.

THE ISSUE

Evolution of communication issues within the controls industry is the subject of a complete paper, and several of the authors in this edition cover various aspects of that topic. Though a detailed focus on control evolution is not appropriate here, a brief discussion of some key industry concerns is important to highlight the urgency of this issue. During the mid to late 1980s, end users had concerns regarding communication with control systems. Trends in the industry were towards system integration, distributed direct digital control, user friendly interface, personal computer front ends, and flexible systems with ease of use as a goal of the system. The communication discussion involved each of these trends. These trends became more complicated, end users noted, when more than one manufacturer's control equipment was installed in the same building. This meant that there was more than one front end system, data could not be shared between systems, and control could not be integrated among systems. Because of these complications, many end users did not find their systems easy to use and the issue of an open protocol developed.

It was also believed that an open or standard protocol, employed in the design of all new control systems, would allow end users to mix and match various manufacturers' components in the same system. This expectation, though possible to achieve with a standard communication protocol, is much more a function of the control sequences executed by controllers. As a result, the mere existence of a standard is not likely to address this desire. Unfortunately, even today there is still great confusion over this issue.

To expand, it is key that two of the control trends above were central themes in the call for open protocols. At the core of the open protocol issue are end user requests for:

• Remote communication from a front end to more than one system
• Standardized networking for communication between distributed controllers

It is important to read these two bulleted items again, because there is a great deal of uncertainty in the industry about the distinctions that must be drawn between these two issues.

CONNECTIVITY AND INTEROPERABILITY

This author coined the term "link" about 15 years ago, to refer to the first communication issue. This term referred to the need for a standard interface between personal computer (PC) front ends and control systems. PCs were being used extensively for communication with control systems of multiple manufacturers, and software interface protocol end was proprietary. Use of a common protocol or communication "link" between a PC and multiple systems was posed as one solution to this issue. The author later referred to this as a "connectivity" solution, but the complexity of the issue demanded more consideration.

The second critical issue that was raised by users and noted above was the desire to be able to easily upgrade systems and to do so with multiple manufacturers' devices. In fact, there was a segment of the industry that wanted fully interchangeable controllers, from any manufacturer, to be able to function in the same DDC system. Today we would refer to this as full-scale "interoperability." In fact, there is a growing consensus in the industry that there are levels of interoperability from simple interface at the low level, to complete interchangeability at the high level.

It is clear that interoperability is an essential element of any DDC system. Consider the growth of distributed DDC, which was made possible by technology developments in the controls, electronics, and data processing fields. Distributed DDC systems are cost-effective on individual pieces of equipment and have become common on devices as small as VAV boxes. Key to the use of these controllers with a complete system, however, is the ability to provide a means for network communication between these controllers, other control products, and a front end, typically a PC.

WHAT IS A PROTOCOL?

Knowing drivers for development of a communication standard is helpful, but is anyone still unclear on the definition of a protocol? In the simplest of terms, a protocol is a set of rules that allows one computer to understand what another one is saying. The key elements of a protocol define:

- Format of the data
- Information necessary for data conversion between machines
- Timing to define the data transmission speed and sequence

Written computer instructions which make up the elements of a protocol are generally called a source code.

A protocol exists wherever two systems must communicate, and historically it was common practice for the protocol source code to be proprietary. Interestingly, this is not unique to the controls industry. The issue may be found in every aspect of the computer industry, and within any industry that integrates computers into products. The search for a solution in other industries led to development of independent research organizations and to significant corporate investment. Regardless of the industry, questions always seems to revolve around the issues of proprietary, standard, and open protocols.

OPEN AND STANDARD PROTOCOLS

As the issue is explored, it quickly becomes evident that the protocol is simply a piece of software that either aids or impedes an owner's ability to meet his needs for the system. It is important to make this statement up front, because no protocol is a solution in itself. Rather, intelligent application of the protocol in the design and development of systems that are intended to meet an owner's need is the solution.

To craft solutions in other industries, research groups came into being that were focused on providing solutions through standard communication between mainframe computers. The Corporation For Open Systems was one such. Another solution, Manufacturing Automation Protocol (MAP) was an early result of a major expenditure made by General Motors Corporation to ensure that all production control systems used standard communication. In each of the above cases, communication guidelines were provided by the International Standards Organization (ISO), which provides a model for developing communication standards.

So, what is the difference between open or standard protocols? Quite simply, **open protocols** differ from other protocols only because the source code is not proprietary, it is published and available, but is often controlled by a company. A **standard protocol**, on the other hand, while also being published and available, is controlled by a standards organization. BACnet is a standard protocol, published by ASHRAE. The source code is published and the intent is that this protocol would be designed into more than one manufacturer's system, and allow for a *standard* in controller communication. Through ASHRAE, the industry focused this effort on meeting the critical needs of the buildings industry. Other open protocols have also been offered to the industry as solutions.

To return to an earlier theme, whether a protocol is open or standard does not address an owner's needs. It is critical to carefully evaluate the needs for a particular project, and to then select one or more systems and protocols that can meet those needs. A final issue regarding standard protocols is that some method of compliance to the standard must be provided, and the status of this issue for BACnet from ASHRAE and LonWorks from Echelon is discussed below. Note that "open system" is yet another term used in the industry and also discussed below.

SYSTEM INTEGRATION OR INTEROPERABILITY

"System integration" is a term that the author has long used to refer to the requirement for coordinating control and other activities such as access, fire, etc., that occur among all the components in a building. In the past, the topic "interoperability" often assumed that there may be more than one level of sophistication or architecture required. More complex systems have traditionally required higher level architectures to accommodate their needs, whereas such complexity would overburden simple systems.

DDC systems to date have evolved around architectures that take a hierarchical or peer-to-peer approach. Solving the interoperability needs under that scenario takes one approach; however, a new level of complexity is being proposed by others in the industry and that is to further distribute system intelligence. This approach would involve multiple individual components, such as schedulers, smart sensors, and PID loop controllers, that reside on a flat architecture.

Given the increasing array of options, it remains as critical for owners to define requirements prior to making any system purchasing decisions. **Interface interoperability** may be viewed as a communication issue, meaning that different systems may be connected and share data. **Control or interchangeable interoperability** is more focused on the idea of controller integration, which mixes more than one manufacturer's controller within a system. The key, of course, is that the controllers must operate as though they were designed to be a system, again a result that is not ensured, simply based upon standard communication. At the system level, multiple complete systems are integrated. In most cases, the system level integration does not integrate controller level functions. Rather, it uses a single front end for programming, monitoring, and other PC functions with all the systems. This type of integration is very similar to a gateway which is discussed below, under "implementations for the future."

Open systems or integration introduces a number of confusing variables into the discussion of standard protocols. Some of the most critical concerns include: warranty, service, maintenance liability, and control integration. Warranty is a question that arises with these systems because each manufacturer would be hard-pressed to identify legitimate warranty claims. Legitimate claims would be those involving traditional problems that could not be blamed on other controller interference, design error, field installation problems, etc. Service and maintenance liability are similar to the warranty issue; however the key here is, who does the owner call for service? Further, the challenge is to ensure that unnecessary site visits and finger pointing do not result in extended downtime.

Control interoperabilty is the last and perhaps the most critical problem. In order for these controllers to work as a system, the designer and installer must plan for control interaction. This means that the control loop in one device could overrule the internal design algorithm in a second device. This is extremely dangerous, particularly where the second device does not have sufficient data to provide effective control.

These problems present significant obstacles to system interoperability. Options for developing such systems cannot be addressed until the specific requirements for the system and the environment where it will be installed are resolved.

STANDARDS IMPLEMENTATION OPTIONS

The above introduction to industry trends and to protocols is limited, but it provides a framework for discussing where the industry is headed. There have been any number of short- and long-term solutions posed to address industry concern over the number of proprietary packages in use for networking and communication. For the most part, these proposed solutions fit into the three categories below, and each will be discussed.

1. Alliances: a hardware or software gateway package to translate between protocols
2. Industry network standardization on an existing protocol such as LonWorks, or
3. DDC network standardization on a new protocol such as BACnet

ALLIANCES AND GATEWAYS

Alliances between manufacturers who share protocols and offer owners a hardware or software conversion package have been an effective shortcut to solving these issues. Protocol conversion requires cooperation and the development of a device to act as a translator between PC front ends or controllers and control systems. These conversion packages are often called "gateways." With this option, there is no change to the existing protocol. A package is developed that can interpret that protocol, convert it to a protocol which the front-end system understands, and pass it through to the front end. This package may be hardware, software, or a combination.

The desirability of this option is that existing systems could be easily modified to allow communication, and that any front end could conceivably function as a standard. As noted, a gateway would be used for universal front-end technology, but gateways are not limited. Gateways can be integrated into stand-alone hardware or distributed controllers to allow system-wide communication. This option does not assume that all existing protocols are acceptable for the long term, but it has been an effective technique for merging existing and new protocols within the same system.

It is highly likely that gateways will remain common with DDC systems, particularly as it becomes more common to integrate disparate systems, i.e., utility meter databases, with controls. Ultimately, systems that use gateways to integrate existing and new DDC equipment and perhaps add interface to a variety of other computer-based facility equipment can fit the definition of an open system.

EXISTING PROTOCOLS

The challenge facing this industry has appeared to be choosing either a new or existing protocol as a standard for control network and interface communications. The primary distinction is whether there will be one protocol, such as BACnet, or if standardization will occur on two or more protocols. The most likely contender for a second protocol standard at this point is LonTalk®, part of the LonWorks offering.

The LonWorks offering has generated enthusiasm in the industry, and in fact has been used as the basis for a number of control-related product developments. Recent development of the LonMark Interoperability Association, an independent association supported by manufacturers and others, is more evidence of the viability of this option. The key here is that LonWorks is not a communication standard because it is controlled by Echelon, an independent company. Yet, clearly, the LonWorks offering is open and has already been implemented and offers much to the industry. Availability of multiple protocols is a workable solution if owners commit to careful design and development of specifications.

A NEW STANDARD

Developing a state of the protocol for controller networking with a direct component to enable standard remote communication is the option that ASHRAE worked on from 1987 until 1995. The result — BACnet — is a package that was designed to meet current, as well as future, system needs. This option, when combined with gateways, would offer a comprehensive solution. Development of a new protocol, though viable, presents complications as well, because availability of the standard is only the first step. After the standard is available, time must be allowed for products to be developed that apply the standard.

This general coverage does not address many of the basic issues being considered by controls manufacturers and the ASHRAE Standards Committee. The chapters that follow will shed more light on many of these issues.

WHERE THE INDUSTRY IS HEADED

Based upon the drivers and issues that have been outlined in this chapter, options for users include two protocol standards for networking and communication. Though some would say that LonWorks is not a standard, but an open communication protocol controlled by a company, for our purposes it appears to be a de facto standard. Each of the standards, or any option considered, must be evaluated by the user to ensure that it meets requirements for any particular project. Timing and cost must also be considered, because application of a standard may require an investment in time and dollars. As appropriate, users will likely continue to need gateways to utilize and integrate existing technology. These options make it possible to integrate control systems with one front end, thus simplifying use and interface with these systems.

The establishment of a standard protocol is exciting, challenging, and necessary, but will never outweigh a clear and explicit statement of requirements. It is now more important than ever for managers to become conversant with the language of protocols and to track the progress of this effort. This is because options for new and existing systems will be affected by whatever action is taken. It is also important because information is critical to the effective management of systems, and access to data is dramatically affected by this issue. Also, system communication remains the best means of maintaining their controls and ensuring their overall performance, and the topic of protocols cannot be separated from any discussion of system communication.

The best first step that each individual and the industry can take is to understand present and future requirements for a DDC system and how a standard can impact those needs. With this information, it will be possible to make intelligent and effective decisions about the communication products that make sense for our industry.

4 Network Protocols

Viktor Boed

CONTENTS

INTRODUCTION

The most common classification of networks for data, voice, and graphic communications are local area networks (LAN), wide area networks (WAN), and metropolitan area networks (MAN). Due to the size of most facilities (campus, research, production, etc.), LANs seem to prevail in facilities automation applications. LANs became very popular for two main reasons:

- Data sharing
- Resource sharing

From their PC workstations connected to the LAN, users can access other computers connected to the network, files allocated on network computers, or dedicated file servers and drives. LANs (also referred to as "information highways") can be shared by various systems utilized by facilities, such as:

- Office systems (E-mail, word processing, accounting)
- Engineering systems (computerized design and drafting, project management)
- O&M systems (work scheduling, inventory control)
- Real-time systems (building automation, power plant, metering, security, CATV, and other systems)

The condition for coexistence of computerized systems on the same network assumes they all have network-compatible communications protocols.

Network protocols — proprietary, open, and standard — are utilized for interfacing comput-erized systems connected to the network. Standard protocols are based on the OSI model. The majority of industry standard or proprietary protocols adhere to some degree to the structure of the OSI model. Some protocols combine some of the layers of the OSI model, creating so-called **collapsed architecture**.

Examples of such protocols are Xerox Network Systems (XNS), transmission control proto-col/internet protocol (TCP/IP), building automation protocols, such as the Canadian Automated Building (CAB), and ASHRAE's Building Automation Controls Network (BACnet) protocol, and others.

OSI LOWER LAYER PROTOCOLS

ETHERNET (IEEE 802.3)

A significant number of facilities utilize **Ethernet** protocol for their LANs. Also, several BASs have Ethernet interfaces marketed for higher level (controller to controller, or controller to OWS) communications. Ethernet became a symbol for CSMA/CD communications method and is closely associated with IEEE 802.3 standard.

Ethernet uses baseband communications (generally), in a bus or star topology, communicating over coax or fiber optic cables at speeds of 10 Mbps to 100 Mbps. Using a 50-Ω coax cable, Ethernet is good up to 2.5 km (about 1.55 miles), with maximum cable lengths up to 0.5 km (about 0.31 miles) before the use of repeaters or signal generators is necessary. IEEE 802.3 defines the maximum number of 1024 data terminal equipment (DTEs) connected to a single network. However, linking several networks together by using gateways can extend this number.

The following Ethernet variations are common in the industry:

- A **standard Ethernet** originally defined for baseband networks with thick coaxial cables is known as **10BASE5** (**10** Mbps, **BASE**band, **.5**-km maximum cable length in each segment). With the use of repeaters, the maximum cable segment length can be extended to 2.5 km between any two transceivers, called medium attachment units (MAU). The number of nodes per segment is 100.
- A so-called **thin Ethernet** utilizes an RG-58 coax cable and the cheaper BNC-T con-nectors instead of MAUs, and is less expensive and easier to install. The network is known as **10BASE2** (**10** Mbps, **BASE**band, for cable lengths up to **.2** km), with the maximum of 30 nodes per segment, and the maximum of 1 km network length with repeaters.
- An Ethernet using **unshielded twisted pairs** for communications is defined as **10BASE-T** (**10** Mbps, **BASE**band, **T** twisted pair, with a maximum cable length of up to .1 km). The 10BASE-T is used to connect nodes to "central" hubs. Despite the distance limitation of 100 m, the 10BASE-T network is very popular, due to the use of inexpensive cables and RJ-45 connectors. By using hubs (repeaters), the network can be extended to the maximum length of 2.5 km.

Network devices (PCS, controllers, etc., also called DTE in networking) can be connected to the Ethernet via **medium access units (MAU) or transceivers**. There can be a maximum number of 100 such devices on any one segment. MAUs or transceivers contain transmit, receive, and collision-detection circuitry.

A **transceiver cable** connecting a network device to the transceiver can be up to 50 m (about 155 ft) long for 10BASE5 networks.

INTERCONNECTING LAN ETHERNET

Repeaters

Ethernet cable segments are connected to each other by a **repeater** (Figure 4.1). The main function of the repeater is to regenerate the signal from one cable segment to another. Most repeaters connect at the physical layer of the OSI model (Figure 4.2), and do not provide network management functions. In fact, they are completely transparent to the network.

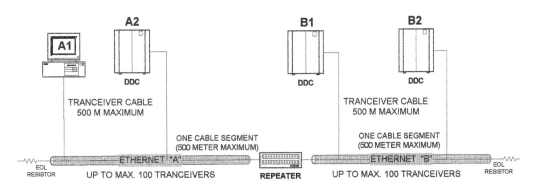

FIGURE 4.1 Use of a repeater on an ethernet.

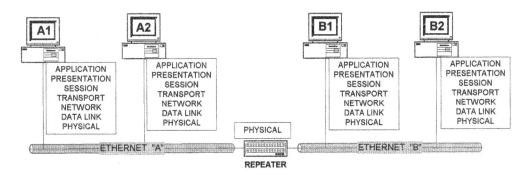

FIGURE 4.2   OSI model layers with a repeater.

Bridges

Bridges interconnect IEEE 802 LANs of similar MAC layers at the data link layer of the OSI model. A bridge (Figure 4.3) monitors traffic on two homogenous networks, and by looking at the source and destination addresses of each packet of data at the MAC sublayer, transmits and/or filters data (ignores packets with designations for the local subnet and forwards packets designated to go to the connecting subnet). Smart bridges can learn the addresses of nodes on different network segments. By using a series of bridges, designers can link a large number of subnetworks together.

A bridge/router or **brouter** is a smart device that combines the characteristics of bridges and routers. Brouters can be used to interconnect multiprotocol networks.

Bridges can be defined to filter out data packets addressed to or from a defined network user, based on the addresses or length of the data packets. Another function of the bridge is to convert, let's say, an Ethernet message to a token-passing message in the case of two networks with different MAC protocols. Since bridges connect on a data link layer, they connect networks regardless of the upper layers of the protocol (Figure 4.4).

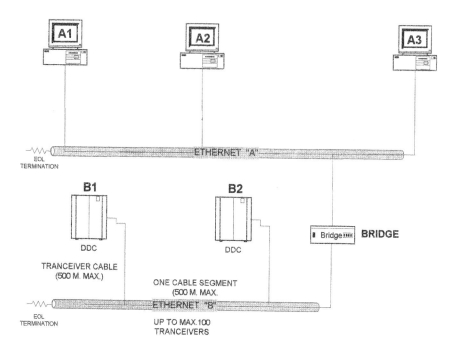

FIGURE 4.3   Bridges between LANs.

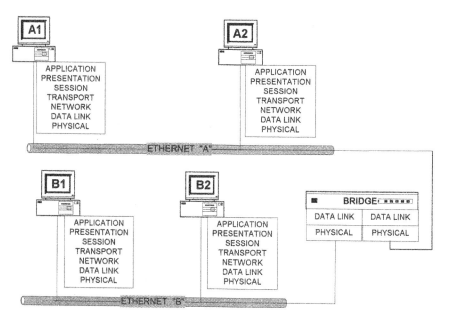

FIGURE 4.4   OSI model with a LAN bridge.

LAN bridges connect systems on the data link layer of the OSI model. The data link layer consists of logical link control (LLC) sublayer and medium access control (MAC) sublayers.

In a network device (computer) OSI model protocol, the data link layer's **LLC sublayer** interfaces to the higher network layer. The bridge LLC sublayer exchanges data units with peer LLC sublayers. The LLC protocol data unit (PDU) consists of addresses of the so-called source service access point (SSAP), as well as destination service access point (DSAP), also defined in the MAC sublayer, along with the control and information fields:

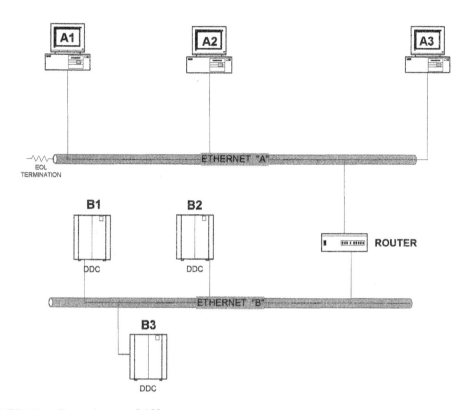

FIGURE 4.5   Router between LANs.

- DSAP
- SSAP
- CONTROL
- INFORMATION

The MAC sublayer defines the frame structure and CSMA/CD procedures.

Routers

Routers are used in network devices supporting a network layer (Figure 4.5). Routers can extend the LAN and reduce the traffic by selecting the most efficient path to the destination by examining the network addresses of each packet, and by routing it the most efficient way.

Since routers work on the network layer, they have access to the routing information. The information is defined either in a routing table of each router or the routers learn the addresses (which pocket goes where). Thus, they are able to increase network throughput by determining the best route for each pocket transmitted on a complex web of subnetworks. Routers provide extension of LANs and reduction of cross-network traffic. Routers are designed for specific network protocols, therefore they can handle only data packets of that protocol (i.e., TCP/IP, OSI, etc.).

Since the network layer provides routing and flow control, routers having access to other computer network layers can provide the most suitable routing in a complex network configuration. The more intelligent routers can also provide network connection multiplexing, error detection, flow control, sequencing, segmentation, and reassembly of data packets. This saves computer resources, since routers can segment larger data packets, assembled by different MAC sublayers, to the maximum size acceptable by a specific LAN (Figure 4.6).

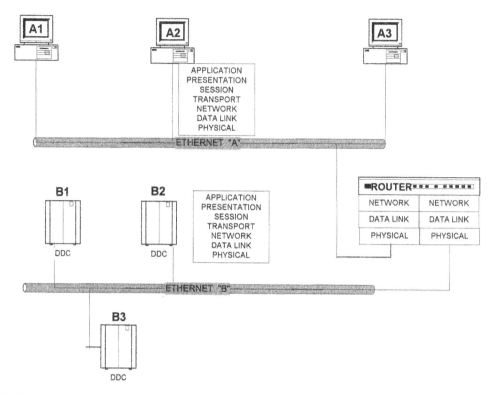

FIGURE 4.6   LAN router in an OSI model.

Gateways

Gateways provide translation of different protocols or heterogeneous networks, allowing interoperability of otherwise incompatible systems and networks (Figure 4.7). Gateways generally operate at the transport or session (higher) levels of the OSI model, affecting their complexity. The amount of processing they are required to perform affects their speed. Gateways are used in the computing as well as the building automation industry to interconnect incompatible networks. Such interconnection can preserve the proprietary nature of subnetworks and at the same time interface the subnetwork to the common (or open) LAN.

Gateways can be half- or full-network nodes. A full gateway is a member of both networks (Figure 4.8); half gateways are linked by an intermediate protocol and communication link.

Gateways on a server are often used in facilities or industrial applications. An example can be a SCADA network server with interfaces to, for example, power plant controls and automation system, building automation network, and to a facilities administrative network.

Ethernet LANs are popular in facilities communications and networking to transmit real-time as well as data processing-related data. Several BASs on the market utilize Ethernet for controller-to-controller or controller-to-OWS communications. A backbone Ethernet can be utilized for systems-to-system (network-to-network) communications as well.

ARCNET

ARCNET was introduced in 1977 as a proprietary protocol and became an ANSI standard in 1992. The ARCNET topology is star or bus or the combination of the two, using active or passive hubs. The communication speed is 2.5 Mbps over 4 miles distance using active hubs. There can be up to 10 active hubs on a network with a maximum number of 255 network nodes (controllers, PCs, etc.).

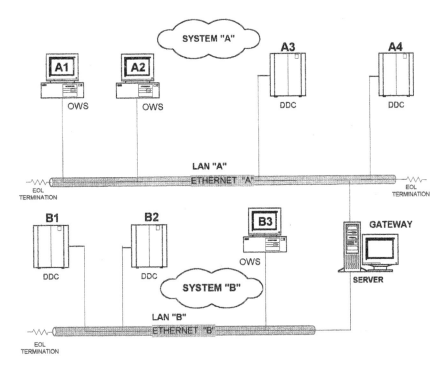

FIGURE 4.7   Gateway between LANs.

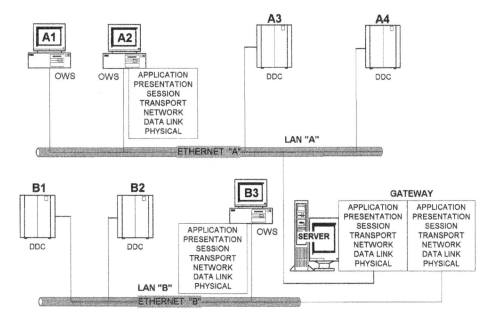

FIGURE 4.8   A LAN full gateway on a server.

The maximum allowable distance between network nodes is 2000 ft. The physical media can be coax cable (RG 62/u), fiber optic cable, or twisted shielded cable (four pairs). ARCNET is a baseband LAN with a deterministic token-passing access mode. The token on a network is passed to the next higher number node (1-255-1) on the network, regardless of its location on the network. The time for a node to receive the token can not only be determined but also guaranteed.

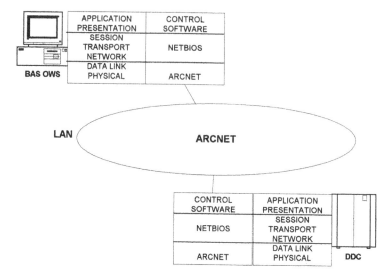

FIGURE 4.9   OSI: ARCNET layers in a LAN application.

ARCNET defines the two lower layers of the OSI model: data link and physical layers.
The token-passing MAC defines five transmission types:

- Invitation to transmit (ITT) — is, in fact, the token
- Free buffer enquiry (FBE) — a query from the transmitting node to destination node to
  check buffer availability
- Data packet (PAC) — is the data transmitted between nodes (8 to 516 characters)
- Positive acknowledgment (ACK) — acknowledgment of recipient
- Negative acknowledgment (NAK) — of FBE and PAC from the destination node

Over the years, ARCNET has developed a large customer base for smaller LAN real-time
automation systems, including building automation DDC and industrial controls systems (Figures 4.9,
4.10). Its success can be attributed to its high speed deterministic nature and its high reliability.

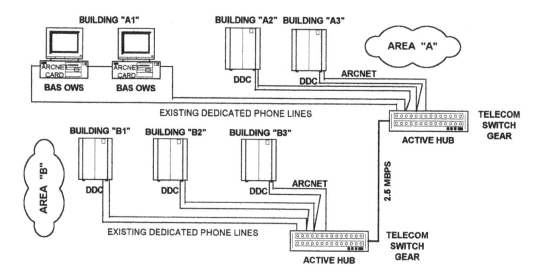

FIGURE 4.10   Example of an ARCNET architecture.

TRANSMISSION CONTROL PROTOCOL/INTERNET PROTOCOL (TCP/IP)

The TCP/IP protocol was developed for the Department of Defense agency's wide-area networks prior to introduction of the OSI model. Later on, TCP/IP was implemented by other users and computer manufacturers and became a standard widely utilized in networking today. TCP/IP is related to transport (TCP) and network (IP) layers of the OSI model, but has utilities for upper layers as well. Some manufacturers use proprietary protocols in session (5), presentation (6), and application (7) layers. Relaying functions between TCP/IP networks is via routers or gateways.

Internet Protocol

The Internet protocol (IP) provides so-called connectionless mode of data transfer over the network (no logical connection is set up before data transmission) (Figure 4.11). The data are transmitted as independent units, called "data grams." The IP provides services, such as data exchange, addresses, and routing, independent of the media and network topology.

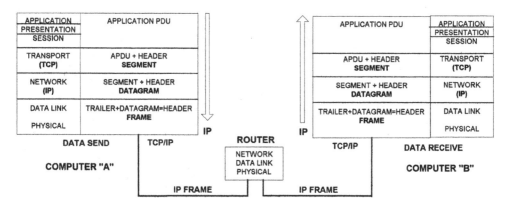

FIGURE 4.11 An OSI IP in a LAN application.

Transmission Control Protocol

The transmission control protocol (TCP) is a connection-oriented protocol (there is a logical connection setup between the transmitting and receiving nodes and the network before data transfer occurs) (Figure 4.12). TCP also provides end-to-end transport services and data integrity support, such as sequencing, flow control, reliability data checks, etc. Another protocol, user datagram

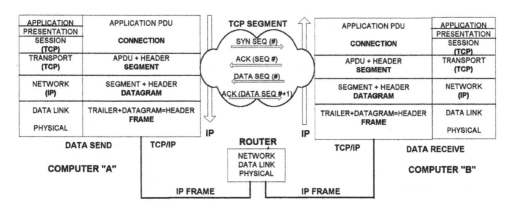

FIGURE 4.12 An OSI TCP in a LAN application.

protocol (UDP), is also available on the same layer. UDP is a connectionless protocol providing routing information with each message, allowing name assignment to each protocol address, and providing for remote network management.

TCP/IP protocols with data rates of 1.5 Mps to 45 Mps are used in the Internet, for example, to provide worldwide electronic mail service (E-mail) (Figure 4.13).

APPLICATION PRESENTATION SESSION (TCP)	TELNET,SMTP,TFTP, FTP,X-WIN,REXEC, DOMAIN,ETC.
TRANSPORT (TCP)	USER DATAGRAM PROTOCOL UDP
NETWORK (IP)	IP GATEWAY PROTOCOLS
DATA LINK PHYSICAL	X.25, WIDE AREA PROT. ETHERNET PROTOCOLS

FIGURE 4.13 TCP/IP protocol suite.

The following is a list of other Internet protocols.

- TELNET: Virtual terminal protocol. All terminals in TELNET are defined as virtual, providing standard means of representation of network terminals, thus relieving the computer from maintaining characteristics of every terminal connected to the network. TELNET also provides remote log in (Rlogin).
- TFTP: Trivial file transfer protocol is a simple file transfer protocol with limited reliability and robustness found in FTP protocols. TFTP supports five PDU types:
  - Read request
  - Write request
  - Data
  - Acknowledgment
  - Error
- FTP: File transfer protocol defines procedures for data (files) transfer between two dissimilar nodes. FTP maintains two logical connections:
  - Control connection between two computers using a protocol interpreter (PI). PI opens, maintains, and closes connections between the two computers.
  - Data management connection called data transfer process (DTP) manages the packeting, sending, receiving, and reassembly of data. The data format can be specified (i.e., ASCII) for file transfer.
- SMTP: Simple mail transfer protocol is widely used for E-mail transfer.
- SLIP: Serial line Internet protocol, and PPP — point-to-point protocol — both for modem connections.

The intention of the short description of the most commonly used network protocols is to aid facilities engineers and managers in understanding the basics of network protocols. Since the area of network protocols is remote from facilities controls and automation and is usually managed by network professionals, readers should always consult their network management whenever contemplating networking of automation systems over common facilities networks. For those interested in standards associated with network protocols, an appendix provides a list of data communications standards, reprinted from *Computer Communications and Networks*, written by John R. Freer, and published by IEEE Press in 1996.

APPENDIX

Data communications standards

EIA standards

RS-232-C (August 1969) RS-232-D (1987)	Interface between Data Terminal Equipment and Data Communication Equipment employing Serial Binary Data Interchange.
RS-422 (April 1975)	Electrical Characteristics of Balanced Voltage Digital Interface Circuits.
RS-423 (April 1975)	Electrical Characteristics of Unbalanced Voltage Digital Interface Circuits.
RS-449 (November 1977)	General-purpose 37-position and 9-position Interface for Data Terminal Equipment and Data Circuit-terminating Equipment employing Serial Binary Data Interchange.

ANSI/IEEE standards

IEEE Std. 488-1978	Standard Digital Interface for Programmable Instrumentation.
ANSI/IEEE Std. 802.3-1985	Carrier Sense Multiple Access with Collision Detection (CSMA/CD): Access method and Physical Layer Specifications.
ANSI/IEEE Std. 802.4-1985	Token Bus Access Method and Physical Layer Specifications.
ANSI/IEEE Std. 802.5-1985	Token Ring Access Method and Physical Layer Specifications.
IEEE 802.6-1987	Metropolitan Area Network Standards.
IEEE 802.7-1989	Broadband Local Area Network Standards.
IEEE 802.8-1987	Fibre-optics Standards.
IEEE 802.9-1993	Integrated Services LAN Interface at the MAC and PHY Layers. (Integrated Voice and Data Network Standard).
IEEE 802.10-1994	Inter-operable LAN/MAN Security.
IEEE 802.11	Wireless LAN Medium Access Control and Physical Layer Specifications.
IEEE 802.12-1994	Demand Priority Access method and Physical layer Specification (100 BASE-VG Anylan).
ANSI X3T9.5	Fiber Distributed Data Interface (FDDI).
ANSI X3T9.3	Fiberchannel.

ISO local area network standards

ISO/IEC 2382-25:1992	Information technology – Vocabulary – Part 25: Local area networks.
ISO 8802-2:1989	Information processing systems – Local area networks – Part 2: Logical link control.
ISO/IEC 8802-3:1993	Information technology – Local and metropolitan area networks – Carrier sense multiple access with collision detection (CSMA/CD) access method and physical layer specifications = IEEE 802.4.
ISO/IEC 8802-4:1990	Information processing systems – Local area networks – Part 4: Token-passing bus access method and physical layer specifications = IEEE 802.4.
ISO/IEC 8802-5:1992	Information processing systems – Local and metropolitan area networks – Part 5: Token ring access method and physical layer specifications = IEEE 802.5.
ISO 8802-7:1991	Information technology – Local area networks – Part 7: Slotted ring access method and physical layer specification.
ISO 9314-1:1989	Information processing systems – Fibre Distributed Data Interface (FDDI) – Part 1: Token Ring Physical Layer Protocol (PHY).
ISO 9314-2:1989	Information processing systems – Fibre Distributed Data Interface (FDDI) – Part 2: Token Ring Media Access Control (MAC).
ISO 9314-3:1990	Information processing systems – Fibre Distributed Data Interface (FDDI) – Part 3: Physical Layer Medium Dependent (PMD).
ISO/IEC 10038:1993	Information technology – Telecommunications and information exchange between systems – Local area networks – Media access control (MAC) bridges.
ISO/IEC 10039:1991	Information technology – Open systems Interconnection – Local area networks – Medium Access Control (MAC) service definition.
ISO/IEC TR 10178:1992	Information technology – Telecommunications and information exchange between systems – The structure and coding of Logical Link Control addresses in Local Area Networks.

ISO OSI standards

I Physical layer

ISO/IEC TR 9578:1990	Information technology – Communication interface connectors used in local area networks.
ISO/IEC 10022:1990	Information technology – Open Systems Interconnection – Physical Service Definition.
ISO/IEC 10173:1991	Information technology – Integrated Services Digital Network (ISDN) primary access connector at reference points S and T.
ISO/IEC TR 10738:1993	Information technology – Local and metropolitan area networks – Token ring access and physical layer specifications – Recommended practice for use of unshielded twisted pair cable (UTP) for token ring data transmission at 4 Mbit/s.

2 Data link layer

ISO/IEC 3309:1993	Information technology – Telecommunications and information exchange between systems – High-level data link control (HDLC) procedures – Frame structure.
ISO/IEC 4335:1993	Information technology – Telecommunications and information exchange between systems – High-level data link control (HDLC) procedures – Elements of procedures.
ISO/IEC 8886:1992	Information technology – Telecommunications and information exchange between systems – Data link service definition for Open Systems Interconnection.
ISO/IEC 10732:1993	Information technology – Use of X.25 Packet Layer Protocol to provide the OSI connection-mode Network Service over the telephone network.
ISO/IEC TR 10735:1993	Information technology – Telecommunications and information exchange between systems – Standard Group MAC Addresses.

3 Network layer

ISO/IEC 8208:1990	Information technology – Data communications – X.25 Packet-layer Protocol for Data Terminal Equipment.
ISO/IEC 8348:1993	Information technology – Open Systems Interconnection – Network Service Definition.
ISO/IEC 8348/DAD1	Addendum to the network service definition covering connectionless-mode transmission.
ISO/IEC8348/DAD2	Addendum to the network service definition covering network layer addressing.
ISO/8472:1988	Information processing systems – Data communications Connectionless-mode network service definition.
ISO/8473:1988	Information processing systems – Data communications – Protocol for providing the connectionless-mode network service.
ISO 8648:1988	Information processing systems – Open Systems Interconnection – Internal organization of the network layer.
ISO/IEC 8880:1990	Information technology – Telecommunications and information exchange between systems – Protocol combinations to provide and support the OSI Network Service.

4 Transport layer

ISO 8072 :1986	Information processing systems – Open Systems Interconnection – Transport service definition.
ISO 8072 Addendum 1 1986	Addendum to the transport service definition covering connectionless-mode transmission.
ISO/IEC 8073:1992	Information technology – Telecommunications and information exchange between systems – Open Systems Interconnection – Protocol for providing the connection-mode transport service.
ISO 8602:1987	Information processing systems – Open Systems Interconnection – Protocol for providing the connectionless-mode transport service.

5 Session layer

ISO 8326:1987   Information processing systems – Open Systems
                                 Interconnection – Basic connection oriented session
                                 service definition.

ISO 8326 Amendment 4:1992 Addendum to ISO 8326 covering session symmetric
                                 synchronization for the session service.

ISO 8327:1987   Information processing systems – Open Systems
                                 Interconnection – Basic Connection oriented session
                                 protocol specification.

ISO 8327 Amendment 3:1992 Addendum to ISO 8327 covering session synchroni-
                                 zation for the session protocol.

6 Presentation layer

ISO 8822:1988   Information processing systems – Open Systems
                                 Interconnection – Connection-oriented presentation
                                 service definition.

ISO 8823:1988   Information processing systems – Open Systems
                                 Interconnection – Connection-oriented presentation
                                 protocol specification.

7 Application layer

ISO/IEC 7498-4:1989   Information processing systems – Open Systems
                                 Interconnection – Basic Reference Model –
                                 Part 4: Management framework.

ISO 8211   Specification for a data descriptive file for informa-
                                 tion exchange.

ISO 8571:1988   Information processing systems – Open Systems
                                 Interconnection – File transfer, Access and Manage-
                                 ment (FTAM).
                                 Part 1: 1988 General Introduction.
                                 Part 2: 1988 Virtual Filestore Definition.
                                 Part 3: 1988 File Service Definition.
                                 Part 4: 1988 File Protocol Specification.
                                 Part 5: 1990 Protocol Implementation Conformance
                                 Statement Proforma.

ISO 8649:1988   Information processing systems – Open Systems
                                 Interconnection – Service definition for the Associa-
                                 tion Control Service Element (ACSE).

ISO 8650:1988   Information processing systems – Open Systems
                                 Interconnection – Protocol specification for the
                                 Association Control Service Element (ACSE).

ISO/IEC 8831:1992   Information processing systems – Open Systems
                                 Interconnection – Job transfer and manipulation
                                 (JTAM) concepts and services.

ISO/IEC 8832:1992   Information processing systems – Open Systems
                                 Interconnection – Specification of the Basic Class and
                                 Full Protocol for job transfer and manipulation.

ISO 9040:1990   Information processing systems – Open Systems
                                 Interconnection – Virtual Terminal Basic Class
                                 Service.

ISO 9041   Information processing systems – Open Systems
                                 Interconnection – Virtual Terminal Basic Class
                                 Protocol –
                                 Part 1: 1990 Specification
                                 Part 2: 1993 Protocol Implementation Conformance
                                 Statement (PICS Proforma).

ISO/IEC 9545:1989	Information processing systems – Open Systems Interconnection – Application Layer Structure.
ISO/IEC 9594:1990	Information processing systems – Open Systems Interconnection – The Directory – Part 1: Overview of concepts, models and services. Part 2: Models. Part 3: Abstract service definition. Part 4: Procedures for distributed operation. Part 5: Protocol specifications. Part 6: Selected attribute types. Part 7: Selected object classes. Part 8: Authentication framework.
ISO/IEC 9595:1991	Information processing systems – Open Systems Interconnection – Common management information service definition.
ISO/IEC 9596	Information processing systems – Open Systems Interconnection – Common management information protocol (CMIP) – Part 1: 1991 Specification. Part 2: 1993 Protocol Implementation Conformance Statement (PICS) proforma.
ISO/IEC 9804:1990	Information processing systems – Open Systems Interconnection – Service definition for the Commitment, Concurrency and Recovery service element.
ISO/IEC 9805:1990	Information processing systems – Open Systems Interconnection – Protocol specification for the Commitment, Concurrency and Recovery service element.
ISO/IEC 10026:1992	Information processing systems – Open Systems Interconnection – Distributed Transaction Processing – Part 1: OSI TP Model. Part 2: OSI TP Service. Part 3: Protocol specification.
ISO/IEC 10035:1991	Information processing systems – Open Systems Interconnection – Connectionless ACSE protocol specification.
ISO/IEC 10040:1992	Information processing systems – Open Systems Interconnection – Systems management overview.
ISO/IEC 10164	Information processing systems – Open Systems Interconnection – Systems Management – Part 1: 1993 Object Management Function. Part 2: 1993 State Management Function. Part 3: 1993 Attributes for representing relationships. Part 4: 1992 Alarm reporting function. Part 5: 1993 Event Report Management Function. Part 6: 1993 Log control function. Part 7: 1992 Security alarm reporting function. Part 8: 1993 Security audit trail function.
ISO/IEC 10165	Information processing systems – Open Systems Interconnection – Management Information Services – Structure of management information: Part 1: 1993 Management Information Model. Part 2: 1992 Definition of management information. Part 4: 1992 Guidelines for the definition of managed objects.

General

ISO 7498:1984	Information processing systems – Open Systems Interconnection – Basic Reference Model.
ISO 7498:Addendum 1:1987	Connectionless-mode transmission.
ISO 7498-2:1989	Information processing systems – Open Systems Interconnection – Basic Reference Model – Part 2: Security Architecture.
ISO 7498-3:1989	Information processing systems – Open Systems Interconnection – Basic Reference Model – Part 3: Naming and addressing.
ISO/IEC 8824:1990	Information technology – Open Systems Interconnection – Specification of Abstract Syntax Notation One (ASN.I).
ISO/IEC8825:1990	Information technology – Open Systems Interconnection – Specification of Basic encoding rules for Abstract Syntax Notation One (ASN.1).

Other ISO standards

ISO 1745-1975	Information processing – Basic mode control procedure for data communications systems.
ISO 2110-1989	Information technology – Data communication – 25 pole DTE/DCE interface connector and contact number (assignments.
ISO 4902:1989	37-pin and 9-pin DTE/DCE interface connectors and pin assignments.
ISO 4903:1989	15-pin DTE/DCE interface connector and pin assignments.
ISO 6159	High-level data link control procedures – Unbalanced classes of procedures.
ISO 6256	High-level data link control procedures – Balanced class of procedures.

CCITT V series recommendations for data transmission over analog networks

V.1	Equivalence between binary notation symbols and the significant conditions of a two-condition code.
V.2	Power levels for data transmission over telephone lines.
V.3	International alphabet number 5 for transmission of data and messages.
V.4	General structure of signals of the 7-unit code for data and message transmission.
V.10	Electrical characteristics for unbalanced double-current interchange circuits for general use with integrated circuit equipment in the field of data communications. (Equivalent to X.26.)
V.11	Electrical characteristics for balanced double-current interchange circuits for general use with integrated circuit equipment in the field of data communications. (Equivalent to X.27.)
V.13	Answer-back unit simulators.
V.15	Use of acoustic couplers for data transmission.
V.20	Parallel data transmission modems standardized for universal use in the general switched telephone network.

V.21	300 bit/s modem standardized for use in the general switched telephone network.
V.22	Standardization of modulation rates and data-signalling rates for synchronous data transmission in the general switched telephone network.
V.23	600/1200-baud modem standardized for use in the general switched telephone networks.
V.24	List of definitions for interchange circuits between DTE and DCE.
V.25	Automatic calling and/or answering on the general switched telephone network.
V.26	2400 bit/s modem standardized for use on four-wire leased point-to-point circuits.
V.26 bis	2400/1200 bit/s modem standardized for use on the general switched telephone network.
V.27	Modem for data signalling rates up to 4800 bit/s over leased circuits.
V.27 bis	4800 bit/s modems with automatic equalizer standardized for use on leased telephone-type circuits.
V.27 ter	4800/2400 bit/s modems standardized for use in the general switched telephone network.
V.28	Electrical characteristics for unbalanced double-current interchange circuits.
V.29	9600 bit/s modem standardized for use on point-to-point four-wire leased telephone circuits.
V.30	Parallel data transmission systems for universal use on the general switched telephone network.
V.31	Electrical characteristics for contact closure type interface circuits.
V.35	Transmission of 48 kbit/s data using 60 kHz to 108 kHz group band circuits.
V.36	Modems for synchronous data transmission using 60–108 kHz group band circuits.
V.40	Error indication with electromechanical equipment.
V.41	Code-independent error control system.
V.50	Standard limits for transmission quality of data transmission.
V.51	Organization of the maintenance of international telephone-type circuits used for data transmission.
V.52	Characteristics of distortion and error rate measuring apparatus for data transmission.
V.53	Limits for the maintenance of telephone-type circuits used for data transmission.
V.54	Loop test devices for modems.
V.56	Comprehensive tests for modems which use their own interface circuits.
V.57	Comprehensive data test for high data signalling rates.

CCITT X series recommendations for data transmission over public data networks

X.1	International user classes of service in public data networks.
X.2	International user facilities in public data networks.
X.3	Packet assembly/disassembly facility (PAD) in a public data network.
X.4	General structure of signals of International Alphabet No. 5 code for data transmission over public data networks.
X.20	Interface between data terminal equipment and data circuit-terminating equipment for start–stop transmission services on public networks.
X.20 bis	Use on public data networks of data terminal equipments which are

	designed for interfacing to asynchronous duplex V-series modems.
X.21	General-purpose interface between data terminal equipment and data circuit-terminating equipment for synchronous operation on public data networks.
X.21 bis	Use on public data networks of data terminal equipments which are designed for interfacing to synchronous V-series modems.
X.22	Multiplex DTE/DCE interface for user classes 3–6.
X.24	List of definitions of interchange circuits between data terminal equipment and data circuit-terminating equipment on public data networks.
X.25	Interface between data terminal equipment and data circuit terminating equipment for terminals operating in the packet mode on public data networks.
X.26	Electrical characteristics for unbalanced double-current interchange circuits for general use with integrated circuit equipment in the field of data communications. (Equivalent to V.10.)
X.27	Electrical characteristics for balanced double-current interchange circuits for general use with integrated circuit equipment in the field of data communications. (Equivalent to V.11.)
X.28	DTE/DCE interface for a start-stop mode data terminal equipment accessing the packet assembly/disassembly facility (PAD) on a public data network situated in the same country.
X.29	Procedures for exchange of control information and user data between a packet mode DTE and a packet assembly/disassembly facility (PAD).
X.30	Standardization of basic model page-printing machines in accordance with International Alphabet No. 5.
X.31	Characteristics, from the transmission point of view, at the interchange point between data-terminal equipment and data circuit-terminating equipment when a 200-baud start–stop data-terminal equipment in accordance with International Alphabet No. 5 is used.
X.32	Answer-back units for 200-baud start–stop machines in accordance with International Alphabet No. 5.
X.33	Standardization of an international text for the measurement of the margin of start–stop machines in accordance with International Alphabet No. 5.
X.40	Standardization of frequency shift modulated transmission systems for the provision of telegraph and data channels by frequency division of a primary group.
X.50	Fundamental parameters of a multiplexing scheme for the international interface between synchronous data networks.
X.50 bis	Fundamental parameters of a 48 kbit/s user data signalling rate transmission scheme for the international interface between synchronous data networks.
X.51	Fundamental parameters of a multiplexing scheme for the international interface between synchronous data networks using 10-bit envelope structure.
X.51 bis	Fundamental parameters of a 48 kbit/s user data signalling rate transmission scheme for the international interface between synchronous data networks using 10-bit envelope structure.
X.52	Method of encoding anisochronous signals into a synchronous user bearer.
X.53	Number of channels on international multiplex links at 64 kbit/s.
X.60	Common channel signalling for circuit switched data applications.
X.70	Terminal and transit control signalling system for start–stop services on international circuits between asynchronous data networks.
X.71	Decentralized terminal and transit control signalling system on international circuits between synchronous data networks.
X.75	Terminal and transit call control procedure and data transfer system

between packet switched public data networks.

X.80	Interworking of inter-exchange signalling systems for circuit switched data services.
X.87	Principles and procedures for realization of international user facilities in public data networks.
X.92	Hypothetical reference connections for public synchronous data networks.
X.95	Network parameters in public data networks.
X.96	Call progress signals in public data networks.
X.110	Routing principles for international public data services through switched public data networks of the same type.
X.121	International numbering plan for public data networks.
X.130	Call set-up and clear-down times in circuit switched public data networks.
X.132	Grade of service in international data communications over circuit switched public data networks.
X.150	DTE and DCE test loops for public data networks.
X.180	Administrative arrangements for international closed user groups.

CCITT series of recommendations X.400 to X.430 for data communication networks – message handling systems (*Fascicle* V I I I .7)

X.400	Message handling systems: system model-service elements.
X.401	Message handling systems: basic service elements and optional user facilities.
X.408	Message handling systems: encoded information type conversion rules.
X.409	Message handling systems: presentation transfer syntax and notation.
X.410	Message handling systems: remote operations and reliable transfer server.
X.411	Message handling systems: message transfer layer.
X.413	Message handling systems: message store.
X.419	Message handling systems: MHS application protocols.
X.420	Message handling systems: interpersonal messaging user agent layer.
X.430	Message handling systems: access protocol for Teletex terminals.

CCITT I-series recommendations for integrated services digital networks (ISDN) (*Fascicle* III.5)

Part I – General

Section 1: Frame of the I-series recommendations – terminology

I.110	General structure of I-series recommendations.
I.111	Relationship with other recommendations relevant to ISDNs.
I.112	Vocabulary of terms for ISDNs.

Section 2: Description of ISDNs

I.120	Integrated Service Digital Networks (ISDNs).

Section 3: General modelling methods

I.130	Attributes for the characterization of telecommunication services supported by an ISDN and network capabilities of an ISDN.

Part II – Service capabilities

Section 1: Service aspects of ISDNs
I.210   Principles of telecommunication services supported by an ISDN.
I.211   Bearer services supported by an ISDN.
I.212   Teleservices supported by an ISDN.

Part III – Overall network aspects and functions

Section 1: Network functional principles
I.310   ISDN: Network functional principles.

Section 2: Reference models
I.32   ISDN: Protocol reference model.

Section 3: Numbering, addressing and routing
I.330   ISDN numbering and addressing principles.
I.331 (E.164)   Numbering plan for the ISDN era.

Section 4: Connection types
I.340   ISDN connection types.

Part IV – ISDN user–network interfaces

Section 1: ISDN user–network interfaces
I.410   General aspects and principles relating to recommendations on ISDN
             user–network interfaces.
I.411   ISDN user–network interfaces – Reference configurations.
I.412   ISDN user–network interfaces – Interface structures and access
             capabilities.

*Section 2: Application of I-series recommendations to ISDN user–network
interfaces*
I.420   Basic user–network interface.
I.421   Primary rate user–network interface.

Section 3: ISDN user–network interfaces: Layer 1 recommendations
I.430   Basic user–network interface – Layer 1 specification.
I.431   Primary rate user–network interface – Layer 1 specification.

Section 4: ISDN user–network interfaces: Layer 2 recommendations
I.440 (Q.920)   ISDN user–network interface data link layer – General aspects.
I.441 (Q.921)   ISDN user–network interface data link layer specification.

Section 5: ISDN user–network interfaces: Layer 3 recommendations
I.450 (Q.930)   ISDN user–network interface layer 3 – General aspects.
I.451 (Q.931)   ISDN user–network interface layer 3 specification.

Section 6: Multiplexing, rate adaptation and support of existing interfaces
I.460   Multiplexing, rate adaptation and support of existing interfaces.
I.461 (X.30) Support of X.21 and X.21 bis based data terminal equipments (DTEs) by
             an integrated services digital network (ISDN).

I.462 (X.31) Support of packet mode terminal equipment by an ISDN.
I.463 Support of data terminal equipments (DTEs) with V-series type interfaces by an integrated services digital network (ISDN).
I.464 Multiplexing, rate adaptation and support of existing interfaces for restricted 64 kbit/s transfer capability.
I.472 Internetworking services and protocols.

CCITT recommendations for teletex services

F.200 Teletex service.
T.60 Terminal equipment for use in the teletex service.
T.61 International repertoires and coded character sets for the international teletex service.
T.72 Terminal capabilities for mixed mode operation.
T.73 Document interchange protocol for telematic services.
T.90 Teletex requirements for internetworking with telex services.
T.91 Teletex requirements for real-time internetworking with the telex service in a packet switching environment.

CCITT videotex recommendation

F.300 Videotex service.

(Reprinted from John R. Freer: *Computer Communications and Networks*, 2nd ed., IEEE Press, 1996. With permission.)

5 Serial Communications

Viktor Boed

CONTENTS

INTRODUCTION

Serial communications is one of the most popular means of data transmission used in the computer as well as the building automation industry. Serial communication is standardized and is defined by EIA or RS standards. The most frequently used standards for automation systems are reviewed in this chapter.

EIA STANDARD RS-232-C FOR DATA TRANSMISSION

Among the most popular data transmission standards utilized for computer and automation communications is the **EIA Standard RS-232-C** (C stands for revision level). The standard was developed for synchronous and asynchronous serial communications between so-called data terminal equipment — DTE (i.e., PCs, controllers) — and data communications equipment — DCE (i.e., modems). The standard is applicable for communications for up to 100,000 bits per second. RS-232 connections are widely utilized in PCs and office automation, as well as in controls and automation installations. *EIA Standard RS-232 is not a communications protocol.*

    RS-232 signals are divided into the following groups: A: ground, B: data, C: control, D: timing, and S: secondary.

    Each signal is assigned to a pin or lead in the RS-232 connector. The interface connector pin assignment is shown in the table below.

Pin Number	Circuit	Description	Direction
1	AA	Protective ground	N/A
2	BA	Transmitted data	To DCE
3	BB	Received data	From DCE
4	CA	Request to send	To DCE
5	CB	Clear to send	From DCE
6	CC	Data set ready	From DCE
7	AB	Signal ground	N/A
8	CF	Received line signal detector	From DCE
9			
10			
11			
12	SCF	Secondary line signal detector	From DCE
13	SCB	Secondary clear to send	From DCE
14	SBA	Secondary transmitted data	To DCE
15	DB	Transmission signal timing	From DCE
16	SBB	Secondary received data	To DCE
17	DD	Receiver signal timing	From DCE
18			
19	SCA	Secondary request to send	To DCE
20	CD	Data terminal ready	To DCE
21	CG	Signal quality detector	From DCE
22	CE	Ring indicator	From DCE
23	CH/CI	Data signal rate selector	To DCE from DCE
24	DA	Transmit signal timing	To DCE
25			

The RS-232 transmission uses unbalanced circuits with voltages between 3 and 25 V. The positive signal to a common reference voltage is interpreted as binary "0"; negative signal as binary "1". The DTE, a computer or controller sends the transmitted data (TD) as a signal (voltage) within a range of +/– 3 to 25 V. This voltage appears between pins 2 (TD) and 7 (signal ground, SG). When the line is idle, the voltage is negative (–), alternating to positive (+) when data are sent. Pin 7, the signal ground, is important in RS-232 data transmission, since it is used as a reference to determine the logical state of the signal.

Due to the nature of the unbalanced circuit and relatively high voltage, the RS-232 transmission is susceptible to noise and crosstalk. It is also susceptible to ground loops as a result of possible voltage differences at different points of the line, caused by different ground potentials. RS-232 is therefore more suitable for point-to-point communications with maximum cable length of 50 ft and data speed of up to 20 kbps.

RS-232 transmission may be synchronous or asynchronous. Communications may be with or without modems. If the design criterion calls for longer distances and multipoint communications, as most automation applications do, serial converters convert the RS-232 signal to RS-485 or -422 transmission.

APPLICATIONS OF RS-232

The application of the RS-232 data transmission standard has many variations and associations with different computing and communications hardware. An application engineer should always consult the vendor's literature to fully understand the details as they relate to the relevant system.

The following is a generic description of the use of RS-232 in the most common communications environments and configurations.

Asynchronous Data Communications over Dedicated Lines

An asynchronous environment is characterized by transmission of character bits enclosed by a start and stop bit (0s and 1s). These are called synchronizing signals. Their function is to alert the receiver on the incoming data, and to give a receiver time to provide certain timing functions prior to arrival of the next data. Transmission timing in asynchronous mode is character based. Asynchronous communication is sometimes characterized by an odd or even number of 1s, checked by a so-called parity check. The communication can take place in a half-duplex mode (over two wires alternating directions), or full-duplex mode (three- or four-wire simultaneous bidirectional transmission) at designated transmission speed — bits per second (bps, or baud rate).

The transmission speed over dedicated lines for local direct connections is 19.2 kbps; for modem communications the speed is usually 1800 bps. Modems operating at higher speed are designed with complex signal modulation and data handling.

Asynchronous communication is inexpensive because synchronization occurs between transmitting and receiving devices on per character basis. The timing tolerances are looser, meaning less expensive DTE and DCE devices. Inaccuracies in data transmission are corrected on per character basis. Error checking is by even or odd parity check, depending on the total number of bits (including the start and parity bits) adding up to even or odd number of logical "1"s. Due to the above characteristics of asynchronous communications, they are frequently used to communicate from PCs to their connected ASCII terminals.

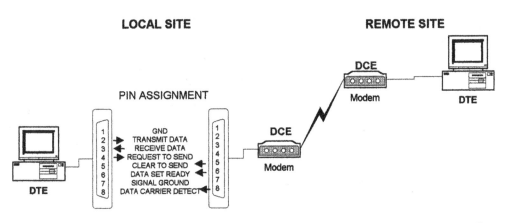

FIGURE 5.1 RS-232 pin assignment for asynchronous data communications over dedicated phone lines.

For modem communications (beyond the maximum cable length), the modems and lines must be ready prior to sending data over an RS-232 connection (Figure 5.1). This is tested out by **data set ready** (DSR) signal generated by modems (DCEs) on pin #6 (1 = OK; 0 = fault or testing mode). If the DSR signal is OK, the computer (DTE) sends out a **ready to send** signal (RTS) on pin #4. The RTS signal is followed by the **clear to send** (CTS) signal originated by the modems (DCE) on pin #5 and by **data carrier detect** signal on pin #8. The **transmitted data** (TD) is sent on pin #2 by the originating computer (DTE); **received data** (RD) from the modem (DCE) comes to pin #3 of the receiving DTE. Pin #1 is the (equipment) **ground**; pin #7 is the **signal ground** (common ground reference for all pins but #1).

Asynchronous Data Communications over Dial-Up Modems

The connection scheme is slightly more complex using a **dial-up modem** (Figure 5.2). The function of a modem (signal modulator/demodulator) is enhanced beyond converting ASCII characters into pulses or frequencies by the dial-up requirement. Since most systems offer auto dial-up capabilities, the modems must first establish phone connection. Upon dialing up the phone number, the **ring indicator** (RI) is set on the remote terminal's pin #22. If the remote **data terminal** is **ready** — indication on pin #20 — the remote device answers the call. If there is no signal detected on pin #20, no communication path is established and a time-out of the connection occurs. When both signals are detected, the call is automatically answered. Following the **auto-answer**, the **data set ready** (DSR) signal is generated by the modem (DCE) on pin #6 (1 = OK; 0 = fault or testing mode). If the DSR signal is OK, the computer (DTE) sends out a **ready to send** signal (RTS) on pin #4. The RTS signal is followed by the **clear to send** (CTS) signal originated by the modems (DCE) on pin #5 and by **data carrier detect** (DCD) signal on pin #8. The **transmitted data** (TD) is sent on pin #2 by the originating computer (DTE); **received data** (RD) from the modem (DCE) arrives to pin #3 of the receiving DTE.

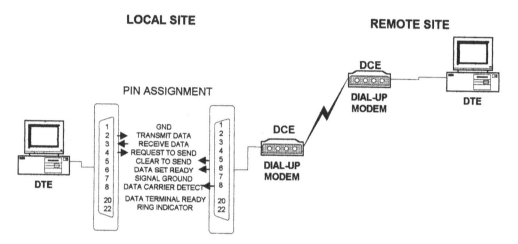

FIGURE 5.2   RS-232 pin assignments for asynchronous dial-up modem communications.

SYNCHRONOUS DATA COMMUNICATIONS

Synchronous data communications were established to increase the volume and speed of data sent over the network. In synchronous data communications, characters to be sent are grouped into logical groups by the computer, called *buffering*, and sent over the network (Figure 5.3). The assembled data strings have so-called synchronization characters (SYNC characters) at the beginning of the assembled block of data. SYNC characters have a unique code to distinguish them from the regular data. Timing for transmission is provided by the DTE (computer) and/or (DCE) modem or by a separate timing signal included in the data string.

Common speeds of synchronous RS-232 connection are 1, 4, 5, 10, and 20 Mbps. Use of dial-up modems usually reduces the communication speed.

Transmission errors are checked by several error detection methods, such as

- *Parity check*, of each character (vertical redundancy check), or block of characters (horizontal redundancy check)
- *Cyclic redundancy check* (CRC), a very efficient method in synchronous communications (also used successfully in LAN communications); CRC is generated by a dedicated chip, shift register feedback or by a software
- And other error or block checks

The transmit timing is provided either by the computer or modem. The **internal transmit timing** (provided by the computer) is on pin #24. **External transmit timing** provided by the

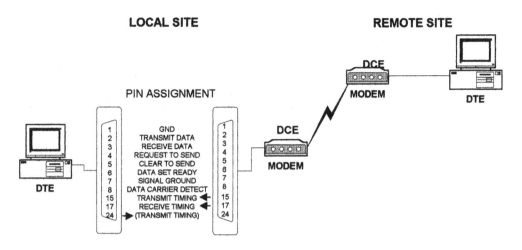

FIGURE 5.3 RS-232 pin assignments for synchronous data communications.

modem is on pin #15. Synchronization of timing of transmitted data (on pins #24 or 15) with timing of the received data (**receive timing**) can be either by the receiving computer or by the modem. The **receive timing** is on pin #17. The receive timing (pin #17) at one end can be looped to the transmit timing (pin #15) at the receiving modem to have a single timing source.

SECONDARY SIGNALS

Besides the primary signals, secondary signals may optionally be utilized in RS-232 communications. They are either used as backup of the primary lines, verification of transmission, or up to provide feedback and data control information to the connected DTEs.

The following secondary signals are incorporated in the RS-232 interface:

Pin #	Description	Abbreviation
12	Secondary carrier detect	SDCD
13	Secondary clear to send	SCTS
14	Secondary transmitted data	STD
16	Secondary received data	SRD
19	Secondary request to send	SRTS

NULL-MODEM CABLES

Standard RS-232 was designed to interconnect DTEs with DCEs. However, it also can be used to interconnect computers or controllers (DTEs) on short local networks (less than 50 ft) without implementing communication modems or serial converters (DCEs).

Since some signals are originated by DCEs (modems), in their absence they have to be taken from connected DTEs. This is being done by so-called **null-modem cables** with cross-connections of pins in asynchronous mode of communication (Figure 5.4). In the absence of synchronous modems in synchronous mode of communications, timing must be provided by external **synchronous null-modem devices**.

Due to distance limitation of the RS-232 data transmission, to connect remote computers or controllers without modems, the RS-232 transmission has to be converted into other forms of serial transmissions, such as RS-422 or the widely used RS-485 data transmission standard. This is being done by so-called serial converters installed at both ends of the data transmission line (more about converters in the following discussions).

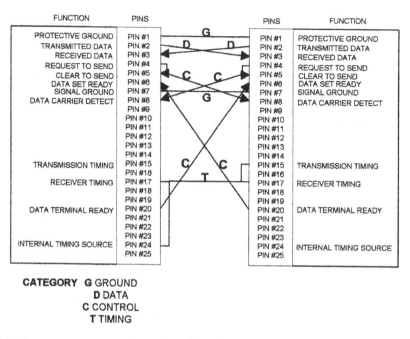

CATEGORY **G** GROUND
          D DATA
          C CONTROL
          T TIMING

FIGURE 5.4   Common cross-connections of a null-modem.

BREAK-OUT BOX

One of the nice features of RS-232 is its simplicity and use of low-cost devices, such as the break-out box, used to test the RS-232 signal conditions (Figure 5.5). The break-out box is an invaluable tool for installation and systems commissioning. The box can be used for cross-connections, monitoring signal conditions, pin testing, and other purposes for testing and diagnostics.

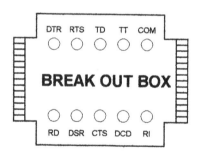

FIGURE 5.5   RS-232 break-out box.

SPECIFICATION OF RS-232 DATA TRANSMISSION

The following are some of the basic parameters for RS-232 data transmission and its serial communications driver.

Driver output voltage for open circuit	25 Vmax
Driver output voltage range	5–15 V
Driver output power off resistance	300 Ω
Receiver input resistance for 5–15 V	3000–7000 Ω
Receiver input threshold	–3 to +3 V

Logical 1	–3 V
Logical 0	+3 V
Maximum length	50 ft
Maximum speed	20 kbaud

IMPROVED SERIAL COMMUNICATION STANDARDS RS-422, -423, AND -449

New EIA standards RS-422, -423 and -449 were developed in the 1970s to improve the limitations of the RS-232 standard. The main improvements are related to the communication speed, interference, and cable length. These new standards define the electrical characteristics for digital circuits for balanced interfaces (RS-422) and unbalanced interfaces (RS-423), respectively (Figure 5.6). Interfaces based on these standards provide communication to devices at greater distances and higher signaling rates.

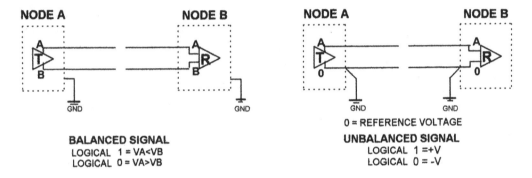

BALANCED SIGNAL
LOGICAL 1 = VA<VB
LOGICAL 0 = VA>VB

0 = REFERENCE VOLTAGE
UNBALANCED SIGNAL
LOGICAL 1 = +V
LOGICAL 0 = -V

FIGURE 5.6   Balanced and unbalanced interfaces.

The basic difference between unbalanced and balanced interfaces is that while the unbalanced interface uses the line voltage in reference to the reference voltage (signal ground), the balanced interface compares one lone voltage to another one to determine the logical state of the signal.

The most common error in unbalanced interfaces is caused by the so-called "ground loop," which is a result of varying ground reference voltages at the node locations. In real life, the nodes (computers, controllers, etc.), are located at different floors of a building, or different buildings of a facility. This could cause differences in ground potentials of several volts between distant signal grounds, resulting in corruption of transmitted data.

Balanced interfaces have reduced error rates due to equal voltage on both wires; they also have reduced noise and interference. This results in higher transmission speeds, lower error rates, and signal transmission over greater distances.

EIA standard **RS-422** describes the "Electrical Characteristics of Balanced-Voltage Digital Interface Circuits." The maximum transmission speed is up to 10 Mbaud for a cable length of about 30 ft, or up to 100 kbaud for a maximum cable length of 4000 ft. The maximum common mode signal voltage is –7 to +7 V; the output voltage for open circuit is less then 6 V, and the minimum differential voltage of the receiver input is 200 mV. Due to the characteristics of the RS-422 interface, they are also used for higher speed baseband data communications.

EIA standard **RS-423** describes the "Electrical Characteristics of Unbalanced Voltage Digital Interface Circuits." The RS-423 standard provides improvements over the RS-232 standard. The maximum transmission speed is up to 100 kbaud for a cable length of about 31 ft, and 3 kbaud for the maximum cable length of 4000 ft. The output voltage for open circuit is 4 to 6 V, which reduces the possibility of crosstalks in the communication circuits. The distance over the RS-232 is improved due to increased sensitivity of the receiver, with the minimum differential input voltage of 200 mV.

The EIA standard **RS-449** specifies a D-series connector with 37 pins for the primary channel, and an additional 9-pin connector for the secondary channel.

The comparison between the pin assignments of the RS-232 and RS-499 is shown in the following table.

			RS-449			RS-232-C	
				AA	Signal ground		GRND
	CM	SG	Signal ground	AB	Signal ground		GND
		SC	Send common				
		RC	Receive common				
	CO	IS	Terminal in service	CE	Ring indicator		CO
		IC	Incoming call	CD	Data terminal ready		
		TR	Terminal ready	CC	Data set ready		
		DM	Data mode				
PRI	DA	SD	Send data	BA	Transmitted data		DA
		RD	Receive data	BB	Received data		
PRI	TI	TT	Terminal timing	DA	DTE signal timing		TI
		ST	Send timing	DB	DCE signal timing		
		RT	Receive timing	DD	Receiver signal timing		
PRI	CO	RS	Request send	CA	Request to send		CO
		CS	Clear to send	CB	Clear to send		
		RR	Receiver ready	CF	Rcv line sig detect.		
		SQ	Signal quality	CG	Sig. quality detect.		
		NS	New signal	CH	DTE sig rate select.		
		SF	Select frequency	CI	DCE sig rate select.		
		SR	Signaling rate selector				
		SI	Signaling rate indicator				
SEC	DA	SSD	Secondary send data	SBA	Sec. transmitted data		
		SRD	Secondary receive data	SBB	Sec. received data		
SEC	CO	SRS	Secondary request to send	SCA	Sec. request to send		
		SCS	Secondary clear to send	SCB	Sec. clear to send		
		SRR	Secondary receiver ready	SCF	Sec. rcv'd line detect.		
	CO	LL	Local loopback				
		RL	Remote loopback				
		TM	Test mode				
	CO	SS	Select standby				
		SB	Standby indicator				

Note: Abbreviations: PRI — primary channel; SEC — secondary channel; CM — common; CO — control; DA — data; TI — timing; GRND — ground.

The parameters listed in the table provide orderly transition from RS-232-C to RS-499 standards. This approach also allows connecting of newer equipment developed with the RS-449 communications standard to existing RS-232-C equipment using appropriate adaptors.

RS-449 provides communications rates of 2 Mbaud, with timing signals up to 4 Mbaud, for a maximum cable length of up to 180 ft, for RS-422 communications. The transmission rate is reduced to 60 kbaud with exponential wave shaping, and 138 kbaud for linear wave shaping, respectively, for RS-423 communication, for a maximum cable length of 180 ft.

Due to the 180-ft maximum cable length, RS-499 has limited use for controls and automation applications.

EIA STANDARD RS-485 FOR DATA TRANSMISSION

The Electronic Industries Association (EIA) has issued the above standard with the prefix "EIA" in 1983 under the name: "Electrical Characteristics of Generators and Receivers for Use in Balanced Digital Multi Point Systems." As the name implies, the standard defines the electrical characteristics for interchange (communications) of binary signals in multipoint interconnection of digital systems. The standard does not specify other characteristics related to interconnection, such as protocol, timing, signal quality, and pin assignments. Therefore, the standard is not an interface standard and should be used in conjunction with other relevant standards, such as the RS standards described in this chapter. The standard is normally applicable for signaling rate of up to 10 Mbps at a common mode voltage range of +12 to –7 V. The electrical parameters are defined for a generator to be able to handle a total of 32 unit loads (defined by current–voltage characteristics of a passive generator and/or receiver) and a total termination resistance of 60 Ω.

The interchange system (communications) includes so-called generators and receivers of communication signals and balanced interconnecting cable with terminating resistors. The only allowed topology is a bus topology. Other configurations cannot be used (consult the manufacturer's recommendations for the device to be interfaced into an RS-485 network).

APPLICATIONS OF RS-485

The RS-485 data transmission standard is widely used for automation system connections. The following are examples of application possibilities utilizing the above standard.

RS-485 Two-Wire Multidrop Network

Up to 32 pairs of generators and receivers (nodes, computers or controllers) can coexist on any one RS-485 balanced transmission network, with a maximum distance not exceeding 4000 ft (Figure 5.7).

Although the above example is called a two-wire network, a third wire for the signal ground is recommended to keep the common mode voltage on the receiver within the range of +12 to –7 V.

Transmission resistors have to be installed at both ends of the network. However, there is no need for transmission resistors at the drop nodes between the two ends. The transmission resistors are usually in the range of 100 to 120 Ω to match the characteristic impedance of the network.

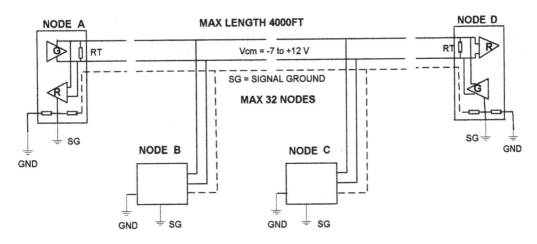

FIGURE 5.7 RS-485 two-wire multidrop network.

The maximum allowable loop resistance is calculated as:

$$R_{loop} = [R_{Term}(1.5 - V_O)]/V_O$$

where R_{Term} is the cable terminating resistance in ohms (generally equal to cable characteristic impedance)

V_O is the minimum voltage that has to be present in the worst case scenario

Based on the calculated loop resistance and the required length, the cable that will meet (or be less than) the calculated loop resistance can be chosen from the manufacturer's data.

Installation of bias resistors may be necessary to force the transmission line into an idle condition, after all nodes have completed their transmission and changed to a receive mode. The bias resistor will force the state of the balanced line to −200 mV on terminals A and B, ensuring the required idle condition (more about different methods of installation of bias resistors later).

RS-485 Four-Wire Network

A four-wire RS-485 network can be used for mixed protocol communication. Such an arrangement allows coexistence of more than one protocol to communicate on the same network. In a four-wire network, one of the nodes is designated as a master and all other nodes are slaves (Figure 5.8). The slave nodes communicate with the master, but not with each other. This arrangement precludes incorrect replies, since the slaves listen only to transmission originated by the master.

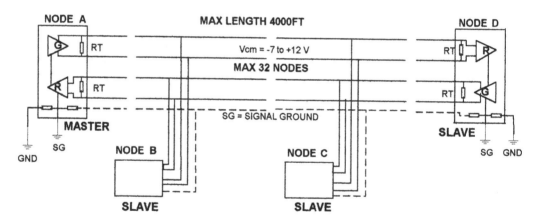

FIGURE 5.8   RS-485 four-wire network.

Biasing an RS-485 Network

As mentioned before, biasing will force the network to idle condition any time the nodes complete their transmission. This can be done by several methods:

- *Single node bias*, using resistors at one of the nodes, thus achieving −200 mV for the idle state. This method will not add unnecessary load to the RS-485 circuit, since the resistors are installed at a single node.
- *Multinode bias*, used when network components (i.e., converters) with internal bias resistors are installed. In such cases, the total bias resistance of the network, which would force the network to idle state, should be calculated.
- *Single node bias* of a low power network termination with high ohm bias resistors. This method is used when the line resistors (RTs) are coupled in both ends of the circuit with capacitance (an AC-coupled termination).

In many automation applications, the RS-485 transmission is used in conjunction with an RS-232 serial communication. The RS-232 signal is converted in a serial converter from RS-232 to RS-485. In such cases the idle state of the RS-485 network can be achieved by:

- Using the RTS signal from the RS-232 converter. A portion of the RTS waveform is used to control the RS-485 driver and receiver. The RTS active signal must be present before the data are sent, and the RTS inactive signal after the data are sent. The timing of the signal is controlled by the software controlling the serial port.
- Using a SD signal to enable the RS-485 driver. A re-trigerable timing circuit may be necessary to assure proper time interval for timing of the transmitted data signal.

Expansion of an RS-485 Network

RS-485 networks can be expanded beyond 32 nodes by using an active RS-485 repeater (Figure 5.9). The repeater listens to the signals at both ends and transmits them to the other sides. The transmission occurs at full voltage level, therefore there can be another 31 RS-485 nodes beyond the repeater connected to the same network. The maximum cable length after the repeater can be up to 4000 ft, in addition to the 4000 ft. before the repeater.

Serial communication is used by many DDC and industrial control systems for data communications on the low end of the system hierarchy. Cost/performance characteristics of this simple data transmission method made it a preferred choice for designers of protocols, integrators, and end users.

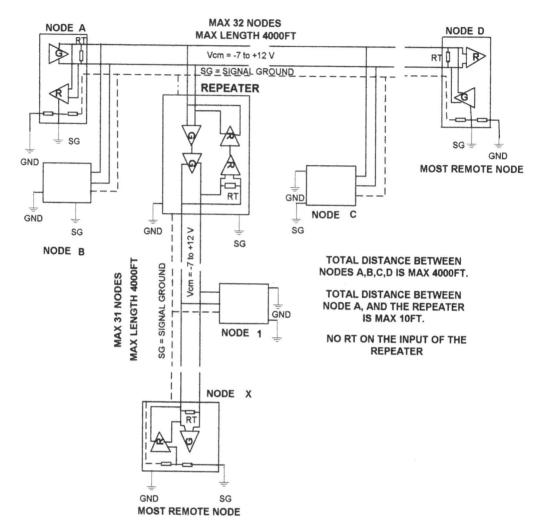

FIGURE 5.9   RS-485 network with a repeater.

6 RS-485 Networks for Facilities Metering at Yale University

Robert Hobbs

CONTENTS

OVERVIEW

We are using RS-485 networks for interfacing remote utility meters located throughout the campus to two of our metering computer servers. RS-485 is used because it allows addressable communications with up to 32 meters on each leg, using a two-wire connection. These meters communicate using MODBUS RTU protocol by means of the Intellution-FIX MBS software driver. The baud rate used throughout is 9600 baud. The maximum distance for each leg is 4000 ft, which was extended on some legs by using dedicated RS-232 modems. See Figure 6.1 for an overview of the network.

Each RS-485 leg uses bus topology. When using the Yale telecom circuits, we do not carry the ground wire. When using dedicated signal cabling, the ground wire is grounded at the source meter only. See grounding diagram in Figure 6.2.

CENTRAL AND SCIENCE AREA NETWORK

The Central Building Utilities Metering System (CBUMS) server accepts inputs from 200 PML-3300 electric meters divided into seven networks. These meters are polled every 1.5 min.

The following is the assignment of meters into individual networks:

Network 1 is a direct connection using a dedicated telephone copper pair leased from Yale Telecommunications (central area). The converter used is a PML COM-128 (4-RS-485 ports) for 43 devices.

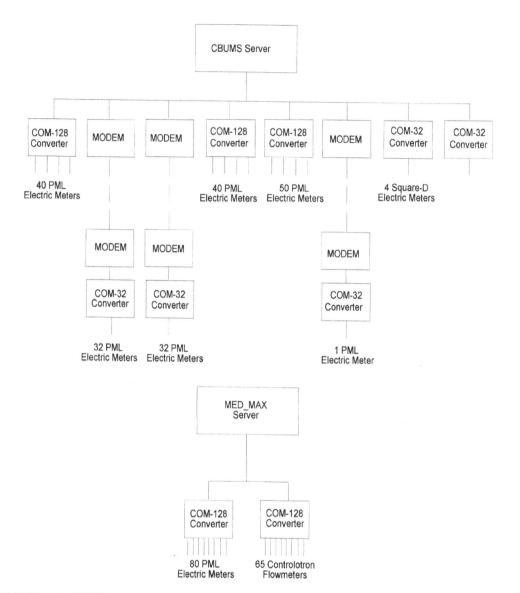

FIGURE 6.1   RS-485 network overview.

Network 2 is a direct connection using a dedicated telephone copper pair leased from Yale Telecommunications (science area). The converter used is a PML COM-32 (1-RS-485 port) for 34 devices.

Network 3 has a modem to Kline Biology Tower (KBT) Telecommunications Hub, and then direct to science and Divinity School areas, using a dedicated telephone copper pair leased from Yale Telecommunications. The converter is a PML COM-128 (4-RS-485 ports) for 46 devices.

Network 4 is a direct connection using a dedicated telephone copper pair leased from Yale Telecommunications (Park St. area). The converter used is a PML COM-32 (1-RS-485 port) for 19 devices.

Network 5 is a direct connection using a dedicated telephone copper pair leased from Yale Telecommunications (Park St. area). The converter is a PML COM-128 (4-RS-485 ports) for 57 devices.

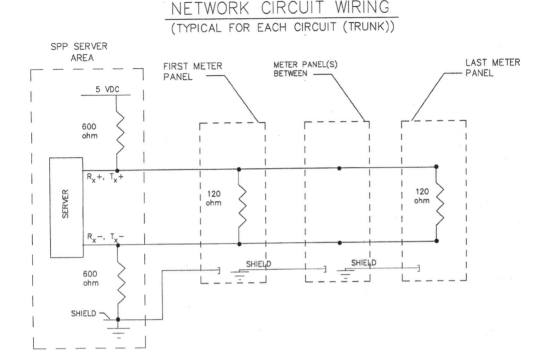

FIGURE 6.2 RS-485 grounding diagram.

Network 6 has a modem to Ingall's Skating Rink using Southern New England Telephone (SNET) dedicated copper pairs. The converter is a PML COM-32 (1-RS-485 port) for 1 device.

Network 7 is a direct connection using a dedicated telephone copper pair leased from Yale Telecommunications (Sterling Memorial Library, SML area). The converter is a PML COM-32 (1-RS-485 port) for 4 devices.

MEDICAL SCHOOL AREA NETWORK

The medical school area server called MED_MAX accepts inputs from 80 PML-7300 electric meters in 8 networks. These meters are polled every 1.5 min. This server also accepts inputs from 65 Controlotron ultrasonic flow meters measuring chill eater (CHW) and condensate in 8 RS-485 networks. These meters are polled every 3 min. All medical school area networks use dedicated copper cables. We use two PML COM-128 converters, each with 4 RS-485 ports for medical area metering, 1 for electric meters, and 1 for chill water and condensate flow meters.

TOPOLOGY

CENTRAL AREA

In the central area, we use existing telecommunications wiring (copper #24 AWG pairs). The telephone center is located in the basement of Woolsey Hall. We located our metering servers in close proximity to minimize RS-485 wiring distances. RS-485 specs recommend total wiring length on each network to be less than 4000 feet. Our baud rate is 9600 bps. Each RS-485 network has 32 or less metering devices.

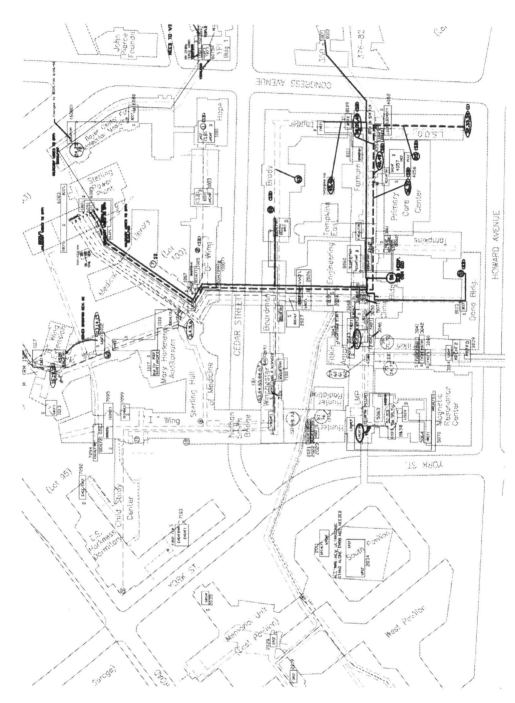

FIGURE 6.3   Medical school area circuit layout.

SCIENCE AREA

Here again we use telecommunications wiring. The telephone center is located in the basement of Kline Biology Tower. KBT is located about 3000 feet north of the central area hub at Woolsey Hall. The science area meters are located less than 4000 feet from the KBT hub. Metering communication between the central and science area hubs is by modem (9600 baud).

MEDICAL AREA

Here we use dedicated cables for RS-485 metering communications. The telecommunications wiring in this area is not as good as in the central and science area, and it was deemed prudent to utilize dedicated wiring here (Figure 6.3).

METER SETUP PROCEDURE

PML

At Yale, we use PML 3300, 7300, and 7330 type electric meters. We use version 2.02 firmware in the latest 7300 meters. For meters having earlier versions, we field upgraded the firmware, using a PML upgrade utility. At 9600 baud, our campus meter standard speed, the upgrade takes approximately 1 hour.

When a version 2.02 meter is installed, we transfer our Yale register assignment (Table 6.1) to the new meter from an existing meter on the same network (cloning). We use "PML Powerview Plus" for the cloning operation. This operation takes approximately 1 hour at 9600 baud. This software also provides a very good diagnostic tool for the PML 7300 network.

For the older PML 3300 meters, we use a DOS program called "MODSCAN" for setup, troubleshooting, and maintenance. When a new electric meter has been set up for operation, it is switched from "ION/PML" mode to "Modbus" mode to interface with our Maxnet system. Our Maxnet system uses Intellution FIX software on the servers. The PML "Powerview Plus" software allows a remote switch from ION to Modbus modes, but not the reverse. PML is working to provide the "reverse switch" capability for our PC servers. Each PML meter is scanned every 90 sec by the PC servers. On our Yale campus metering system, we have the following total number of meters: PML-3300s-197; PML 7300s-80; PML 7330s-2.

CONTROLOTRON

We use Model 990E ultrasonic thermal energy flow meters with an internal Modbus card for communication with our PC servers. Both single and dual channel versions are used to monitor both chill water and condensate flow. Meter registers are set up according to the Yale register assignment table by a Controlotron setup technician. He also programs a unique Modbus address for each meter during setup. Each Controlotron meter is scanned every 3 min by the PC servers running Intellution FIX software with the MBS Modbus driver. For each CHW loop, we monitor either supply or return flow, supply pressure, return pressure, supply temperature, and return temperature. Temperature sensors are 3-wire RTDs brought into the meter on a temperature input card. Pressure sensors are 4 to 20-milliamp looped and brought into the meter as auxiliary inputs. There are 65 Controlotron meters in our medical area.

EMCO

We use EMCO turbine flow meters in the power plants. Each flow meter with temperature and pressure inputs is wired to an FP-93 flow processor. The processed variables are sent from each FP-93 to the server, using a software serial driver developed by Indtech.

TABLE 6.1
Modbus Register Assignment Tables

CONTROLOTRON CONDENSATE METER

LABEL	CH. 1	CH. 2	TAG
SITE NAME	30001	30015	N/A
DATE & TIME STAMP	30005	30019	N/A
LIQ. FLOW AVG.	45005	45018	_CONDAVGF
LIQ. TOTAL	45006	45019	_CONDTOTF
LIQ. SONIC VELOCITY	45007	45020	_RANGEALM
RETURN TEMP	45009	45022	_CONDT
SIGNAL STRENGTH	30011	30025	_ALM
STATUS ALARM	30013	30027	_STATUS
AUX. ALARM BITS	30014	30028	_CRCVR
ANALOG INPUT 1	45012	45025	_SP
ANALOG INPUT 2	45013	45026	_RP

CONTROLOTRON CHW METER

LABEL	CH. 1	CH. 2	TAG
SITE NAME	30001	30015	N/A
DATE & TIME STAMP	30005	30019	N/A
AVG. ENERGY FLOW	45001	45014	_CHWAVGE
TOTAL ENERGY FLOW	45003	45016	_CHWTOTE
AVG. LIQ. FLOW RATE	45004	45017	_CHWAVGLFR
LIQ. TOTAL	45006	45019	_CHWTOTF
LIQ. SONIC VELOCITY	45007	45020	_RANGEALM
SUPPLY TEMP	45008	45021	_CHWST
RETURN TEMP	45009	45022	_CHWRT
DIFF TEMP	45010	45023	_CHWDT
SIGNAL STRENGTH	30011	30025	_ALM
AERATION	30012	30026	_CHWA
STATUS ALARM	30013	30027	_STATUS
ANALOG INPUT	45012	45025	_SP
ANALOG INPUT	45013	45026	_RP

PML 3300 ELECTRIC METER

LABEL	WYE	DELTA	SINGLE PHASE	TAG
AVERAGE VOLTAGE	40008	40004	40004	_VAVG
CURRENT PHASE A	40009	40005	40005	_IA
CURRENT PHASE B	40010	40006	40006	_IB
CURRENT PHASE C	40011	40007	N/A	N/A
KILOWATTS	40013	40009	40008	_KW
POWER FACTOR	40014	40010	40009	_PF
KWH (MSW)	40015	40011	40010	_KWH_A
KWH (LSW)	40016	40012	40011	_KWH_B

PML 7300 ELECTRIC METER

LABEL	WYE	DELTA	SINGLE PHASE	TAG
VOLTAGE PHASE A	40011			_VA
VOLTAGE PHASE B	40012			_VB
VOLTAGE PHASE C	40013			_VC
VOLTAGE AVERAGE	40014			_VAVG
CURRENT PHASE A	40015			_IA
CURRENT PHASE B	40016			_IB
CURRENT PHASE C	40017			_IC
CURRENT AVERAGE	40018			_IAVG
VOLTAGE UNBALANCE	40019			_VALM
CURRENT UNBALANCE	40020			_IALM
POWER FACTOR	40021			_PF
KILOWATTS	40027			_KW
KVA	40028			_KVA
KVAR	40029			_KVAR
THERMAL DEMAND	40030			_TD
MIN THERMAL DEMAND	40031			_TD_MIN
MAX THERMAL DEMAND	40032			_TD_MAX
SLIDING WINDOW DEMAND	40033			_SW
PREDICTED SWD	40034			_SW_PRED
MIN SWD	40035			_SW_MIN
MAX SWD	40036			_SW_MAX
KWH (MOST SIG. BYTE)	40093			_KWH_A
KWH (LEAST SIG. BYTE)	40094			_KWH_B

There is a main FP-93 panel including 36 FP-93s located outside the Sterling Power Plant (SPP) control room. There are also three remote FP-93 panels located at two Yale School of Medicine buildings and in the Sterling Power Plant basement. These remote panels have 12 FP-93s in total. New remote panels are now being ordered from EMCO. They will use RS-485 for communication with our PC servers.

We have a total of 48 EMCO FP-93 turbine flow meters.

Setup in Intellution **FIX** Software

The meter must be set up in the software driver (MBS, OPTO, EMCO). Parameters include port number, baud rate, polling rate, I/O address, register type, and register number.

The meter must be set up in the FIX database. (PP_MAX, MED_MAX, CBUMS_MAX). The following analog input blocks must be set up for each parameter from the meter:

Electric meters: KWH, IA, IB, IC, IAVG, KWD, VA, VB, VC, VAVG, PF, KVA, KVAR
Chill water meters: AVG_LFR, AVGE, TOTE, TOTFLOW, ST, RT, SP, RP, SONIC_VEL, STAT_ALM, AERATION, RANGEALM
Condensate meters: AVG_LFR, TOTFLOW, RT, STAT_ALM, HOURTOT

The meter parameters must be set up in the historical trend modules as well.

The meter points are set up with DDE links in an EXCEL spreadsheet for maintenance and operation purposes.

User screens are set up using the "FIX DRAW" software module. These diagrams are set up in a tree organization and structure. Each screen shows the real-time or calculated point values from a loop overlaid on an applicable flow diagram for ready reference.

7 Case Study: Heterogeneous Interconnected Fieldbuses

Frantisek Zezulka

CONTENTS

INTRODUCTION

The Department of Controls and Automation is part of the School of Electrical Engineering and Computer Sciences of Technical University Brno, Czech Republic. One of our laboratories, part of the department's research and academic programs, is the laboratory of industrial automation. The laboratory utilizes various automation devices such as programmable logic controllers (PLC) of different brand names, industrial electronic controllers, laboratory robots, asynchronous motors with a variable frequency controller, and other laboratory equipment with microcontrollers. All of them have independent control programs and dedicated interfaces; all of them had been connected to their own IBM (or compatible) PCs for programming and operator interfaces.

The principal idea, which started the development of a common serial communication, was to enable interoperability and interface of all these devices on a common network and to create a SCADA-like system in the laboratory. Hence, a decision to utilize common serial communication was made in 1994.

Because of our contacts with colleagues in the U.S., we were informed on LONWORKS® technology, developed by Echelon and supported by Motorola and Toshiba microtechnology earlier than other institutions in Europe. We have decided to apply LONWORKS technology with the LonTalk® protocol as the first backbone segment for our laboratory application.

At the same time, several European serial drivers and protocols for building and industrial controls and data acquisition became available. Expansion of lower-level protocols indicated an increasing demand for interoperability for building and industrial applications. We have continued with integration of our laboratory control devices by selecting other popular serial drivers, such as Profitbus, CAN, the French industrial standard FIP, and the quite new AS-interface (mostly to be applied in logical controllers).

During the last 3 years we have developed gateways between LonTalk and AS-interface. Another interface between Profitbus and CAN was developed, thanks to a grant from the Czech Grant Agency (GACR).

The following configuration has been operational since the end of 1997 (Figure 7.1).

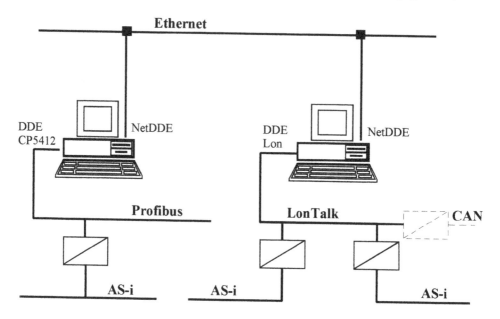

FIGURE 7.1   Layout of the laboratory application.

ARCHITECTURE OF THE LABORATORY APPLICATION

The heterogeneous fieldbus backbone is built from both LonTalk and Profitbus serial communication systems. The physical medium of the Profitbus is a shielded twisted pair; the LonTalk backbone is by twisted pair wires. The philosophy of interconnection of both segments, the LonTalk and Profitbus, will be described in the next paragraph. First, let us describe the individual segments.

LonTalk Segment

The architecture of the LonTalk backbone segment is shown in Figure 7.2. There are several gateways located on the LonTalk backbone segment. The first two are gateways between LonTalk and AS-interface (AS-i), and the next one is a gateway between LonTalk Free Topology / Power Line / Omron. By means of LonNodes cards, several individual devices such as controllers, the variable frequency controller and the control system of the laboratory robot, are connected to the LonTalk segment. Transceivers in Free Topology enable a 78.13 kbps data exchange rate, even though LonWorks technology enables more rapid data translation rate (1.25 Mbps).

The first LonTalk/AS-i gateway consists of two ports RAM and two processors (I8051 and Motorola 68HC11), and is the oldest element of communication in the laboratory application. To connect the gateway to the LonTalk backbone, a proprietary serial member SLTA of LonWorks technology was used. AS-i nodes are formed by standard AS-i active slaves, which enable direct connections of passive sensors and actuators such as switches, valves, and LEDs. One intelligent temperature sensor developed in the department is also connected to the network.

The second LonTalk/AS-i gateway is formed by a standard Neuron® chip and Motorola 68HC11 processor. A powerful parallel interface of the Neuron chip is used to connect with the Motorola microcontroller. The Motorola processor performs host functions of the Neuron chip and supports

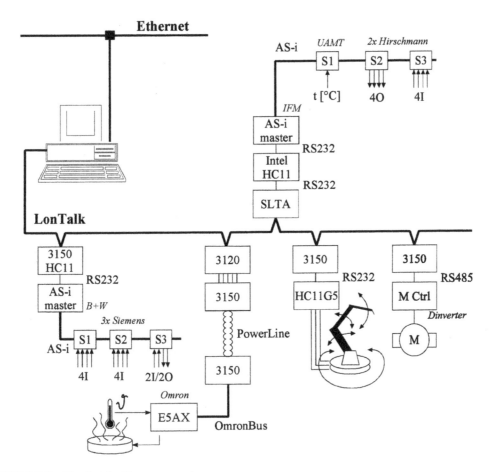

FIGURE 7.2 The first backbone segment.

the RS-232C interface to a conventional AS-i master with a data exchange rate of 57.6 kbps. Active AS-i slaves enable connections of binary switches for general control purposes.

The next connection is formed by a router, also developed in the laboratory. The router is designed with both Neuron 3120 and Neuron 3150 processors. Two Power-Line LonWorks transceivers and an additional Neuron 3150 are utilized for a router. The router enables connection of an adaptive temperature controller Omron E5AX to the LonTalk segment. Communication between E5AX and the router is by RS-232C serial interface. Additional nodes on the LonTalk segment are the variable frequency controller and the laboratory robot.

The variable frequency controller is connected by a universal LonWorks node with a RS-485 interface to the LonTalk segment.

The control box of the laboratory robot is configured as the last LonTalk node. This standard connection via a simple universal LonWorks card enables control of the robot from various nodes of the interconnected fieldbus system.

THE SECOND BACKBONE SEGMENT

This segment is by Profitbus, version Sinec L2. The architecture of the segment is shown in Figure 7.3. There are three active Profitbus stations there: the PC card CP5412, the PLC Simatic S7-300, and the PLC Simatic S5-95. The only passive Profitbus station is formed by a proprietary gateway Profitbus/AS-i. The gateway behaves also as a simple PLC. This device enables control of any AS-i slave by the S5-95, or S7-300, or by the most flexible active station — the PC with a

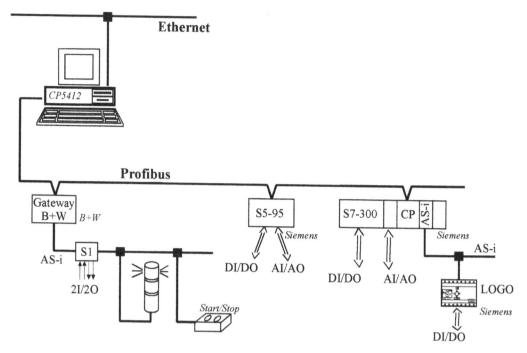

FIGURE 7.3   The second backbone segment.

CP5412 card. The start/stop buttons could control not only devices connected to the Profitbus segment, but also the robot or the variable frequency controller connected to the other backbone segment.

PRINCIPAL GATEWAY BETWEEN BACKBONE SEGMENTS

The gateway between Profitbus and LonTalk protocols is not built by hardware components. The interface software resides in the last gateway. Because of development in the Microsoft® Windows system, the gateway was built by means of DDE (dynamic data exchange) servers. The gateway utilizes two PCs connected to either the LonTalk or the Profitbus segments. A SCADA system (InTouch 5.1) on a MS Windows DDE server with both the CP5412 and the LonTalk master cards also provides an interface to both systems. A Net DDE for an Ethernet connection provides connection of both LonTalk and Profitbus protocols, as shown on Figure 7.1.

EXPERIENCES, PROBLEMS, AND FUTURE OF THE PROJECT

METHODOLOGY OF A FIELDBUS INTERCONNECTION

In general, there are two different ways to realize heterogeneous interconnection of different fieldbus systems:

1. **The first solution** is to use "centralized" interconnection on a higher level by means of PC workstations. The software on this workstation has interfaces to each connected system. Data exchange is provided by these interfaces. This method has been used to provide the principal gateway between backbone segments of the Profitbus and LonTalk.
2. **The second solution** utilizes "decentralized" principles based on a microcontroller card with appropriate fieldbus interfaces and a shared memory. In this case, the messages in the shared memory are visible and accessible from both microcontrollers that manage data from different fieldbus segments.

In comparison, the "centralized" gateway solution is easier and requires less development time, because of a wide choice and utilization of MS products such as DDE servers and OLE (object linking and embedding) on the market. On the other hand, this solution may not be the best for some applications because of the stability of the operating system (MS Windows) and higher cost for utilization of PCs for communication interfaces. The latter was also the reason why we have decided to develop some hardware pieces as gateways in the laboratory.

Another good characteristic of the hardware solution is its higher data translation rate than on PC-based gateways. In the case of the LonTalk to AS-i interface, the critical parameter (bottleneck) for the data translation rate is the serial interface (RS-232C) of the AS-i master. This drawback was remedied in the second LonTalk/AS-i gateway by utilizing the parallel interface of the Lon-Node, which provides more flexibility and higher communication speed than the previous solution.

Poor EMC properties of the variable frequency controller installed in the laboratory is the most serious problem on the LonTalk segment.

We have decided to connect several PLCs of Schneider Electronic (Modicon and Telemecanique) into the laboratory application of interconnected fieldbuses in the future. For these interfaces we will utilize proprietary PLC modules (Modbus module in Telemecanique TSX 37) instead of development of gateways in the laboratory. We will use an Unitelway bus (a proprietary closed industrial protocol) to interconnect the Telemecanique PLCs. Their interconnection to the backbone fieldbus segments will be by a communication driver on a SCADA InTouch system utilizing an OPC software (OLE for Process Control), with OLE and ActiceX methods, instead of DDE on the server. Hence, in the near future, the laboratory industrial communication systems will include six widely used fieldbuses and proprietary lower industrial buses. All systems will be visible and controllable by the InTouch SCADA system.

UTILIZATION OF THE LABORATORY APPLICATION

The laboratory application of interconnected fieldbuses forms an example of how to interconnect different industrial communication buses and interfaces for purposes of interoperability of wide varieties of automation systems. It serves in the academic programs for students in the last year of both masters studies in controls engineering as well as for computer science specialization. Already, several Ph.D. students have used the network for their laboratory experiments with interconnected fieldbuses for their thesis.

Besides the academic programs, the department also conducts courses and seminars to introduce new control and communication methods to the industry. Experts from the industry find examples of how different control systems can be interconnected and can communicate on common communication buses in our laboratory. They can also study and check out the properties of proprietary, open, and industry standard communication protocols used in building and industrial automation applications.

REFERENCES

Benes, P. and Hrdlicka, M., ASI-bus-connecting sensors and actuators, Proc. of EDS '95, Brno, Czech Republic, 1995, pp. 198–200.

Zezulka, F., Final Report of the GA 102/95/1365, UAMT-FEI VUT v Brne, Czech Republic, December 1997.

Zezulka, F., Hrdlicka, M., and Polacek, R., Tools for visualization and process control in heterogeneous fieldbus systems, *Proc. of the Workshop 97*, Prague, Czech Republic, 1996, pp. 221–222.

8 Lower Level Industrial Networks: Fieldbus

Viktor Boed

CONTENTS

INTRODUCTION

There has been a new trend in the development of layered architecture in controls and automation applications since the beginning of the 1980s. This is due, in part, to new (and continuous) development in microelectronics, as well as the desire to provide automation systems to customers for the lowest possible cost. Development in communications and networking, systems engineering, and integration, along with development of new applications engineering techniques, has also played an important role in supporting this trend.

Layered system architecture brings along advantages associated with distributed processing and allows access to data collected at the lowest levels of the systems hierarchy. Lower level networks provide communications on the lowest layers of the systems architecture, *at the controller, transducer*, and *actuator/sensor levels.* They replace the controller-to-sensor current (4 to 20 mA) or voltage loop connections by digital, bidirectional, multidrop serial bus communications. Each Fieldbus-compatible device is enhanced by computing and communications capabilities. Fieldbus-compatible field devices become so-called "smart" devices, capable of executing simple control, diagnostic, and maintenance functions and providing bidirectional serial communication to higher level controllers.

Lower level networks are being widely used in industrial control applications. Even though the main focus of this book is on building controls and automation communications, the information related to this technology can be useful for systems engineers and integrators as well as for building engineers and managers. Some of the technologies described in this chapter are primarily used for specific industrial applications; others are used in specific regions (i.e., in Europe); some of the technologies have already become standards; others became de facto standards, due to the large number of already installed systems. Whether or not these technologies find their way to building automation applications will depend on their marketing by automation vendors and integrators to building owners and on their acceptance by building owners, systems engineers, and integrators.

ADVANTAGES OF LOWER LEVEL NETWORKS

The popularity of lower level industrial networks can be explained by the desire to provide more cost-effective solutions for systems implementation and more processing power on the lower levels of the network. Considering the cost breakdown for an installed system, installation and wiring costs rank among the highest in installations with a large number of field control devices such as controllers, sensors, actuators, and relays — commonly referred to as "field devices" (points). Networks providing communications at the field controller level as well as at the actuator and sensor levels provide cost-saving solutions by considerably reducing wiring cost.

In conventional applications, the field wiring runs from a controller to each field device in a star-like configuration. The configuration of low-end networks is more like a multidrop configuration (points are daisy chained) in the chosen network topology with individual sensors and actuators as nodes residing on the same communication bus (Figure 8.1). Utilizing appropriate low-end networks reduces wiring (material and labor) installation as well as systems check-out cost.

Utilizing communication networks on the lower levels also results in higher reliability of data transmission. In most instances, lower level networks provide the same or similar setup and data-handling features as the more sophisticated local area networks. Due to these features, they represent a step up toward higher level communications at the sensor actuator level. This is also one of their most significant characteristics, distinguishing them from the traditional sensor actuator interfaces, such as individual current or voltage loops. Since the characteristics and setup parameters of field devices on a network are also visible over the network to the connected OWS, commissioning of such systems can be done (at least partly) from the networked OWS or from a display screen of a controller. Ease of commissioning and systems setup results in additional cost saving due to faster completion and startup of the job.

DIVISIONS OF LOWER LEVEL COMMUNICATIONS

While local area networks such as Ethernet or ARCNET are defined for building or factory level communications, lower level networks can be divided into the following categories:

  Actuator/sensor level
  Device level
  Fieldbus level

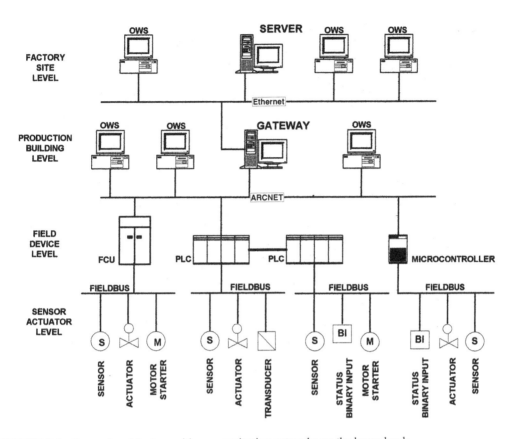

FIGURE 8.1 Layered architecture with communication networks on the lower levels.

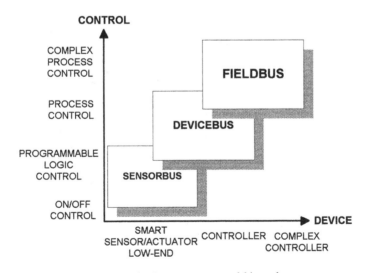

FIGURE 8.2 Lower level communication in the process control hierarchy.

As seen on Figure 8.2, each of these levels is characterized by their hierarchy of process control applications. Similar divisions can be equated to hierarchical levels in building control applications as well.

The main characteristics of the lower level communication networks also can be seen in Table 8.1.

TABLE 8.1
Characteristics of Lower Level Networks

Characteristics	Sensorbus	Devicebus	Fieldbus
Application	Discrete	Discrete	Process
Typical use	Sensor/actuator	PLC	DCS
Data size	< 1 Byte	< 32 Bytes	< 1000 Bytes
Microprocessor	No	Yes	Yes
Intelligence	No	Yes	Yes
Diagnostics	No	Simple	Complex
Response time	< 5 ms	< 5 ms	100 ms
Com. distance	Short	Short	Long
Application	Sensor	Sensor or sensors w/diagnostics	PID w/diagnostics

PROFITBUS

GENERAL

Profitbus is a standard (DIN 19245) used for digital communications on a Fieldbus level. This includes communications on controller levels (i.e., PLC) and also on sensor, actuator, and transmitter levels. The standard evolved from the effort of a collaborative project — Fieldbus — partly funded by the German Federal Ministry of Research and Technology with participation from 13 automation vendors and 5 technical and scientific institutes.

The standard is accepted by hundreds of automation vendor organizations and is used in thousands of applications throughout the world. A Profitbus user organization (PNO) tests conformance to the standard and interoperability of individual products claiming compliance to the standard. PNO is also charged with continuous support and further development of the Profitbus standard.

BASIC CHARACTERISTICS OF THE STANDARD AND ITS ARCHITECTURE

Profitbus devices are either **master** or **slave** devices. Conceptually, Profitbus is a serial Fieldbus, which follows the OSI model layers 1, 2, and 7. Layers 3 to 6 are omitted, since they are not needed for control applications at this level of the network hierarchy.

Layers 1 and 2 are defined in the standard DIN 19245, Part 1, as Fieldbus Data Link (FDL). From this layer, Profitbus interfaces to the physical media using RS-485 transmission standard.

The main characteristics of layers 1 and 2 are

- RS-485 transmission using twisted pair of wires, galvanic separation, shielding is optional
- Maximum line length is 1200 m (3600 ft), extendable to 4800 m (14,400 ft), depending on the transmission rate
- Transmission rate is from 9.6 to 500 kbps
- Number of active and passive stations is 127 max.
- Bit coding is NRZ
- Data transmission is asynchronous, half-duplex
- Profitbus utilizes hybrid bus access

Profitbus uses a modified method, called hybrid medium access for medium access control (MAC) controlling data communication between the masters as well as masters and slaves (Figure 8.3).

The Profitbus MAC assures execution of data transmission by each master within the predefined time interval. Profitbus uses a token passing method among masters. Each master has a right to

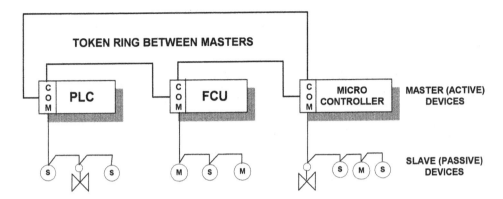

FIGURE 8.3 Hybrid medium access of the Profitbus.

transfer a token within a predefined time interval, assuring maximum token rotation between the master controllers.

The Profitbus MAC also assures data communication between the masters and their connected slaves. The master that owns the token has the right (for a certain time) to communicate with its slave stations (i.e., sensors, actuators, etc.), as well as to pass the token to the next master.

MAC provides other services as well. It recognizes masters and slaves which are either new, turned off, or are faulty, and includes them in or deletes them from data communications. It also detects transmission, addressing, and token-passing errors (i.e., multiple usage, multiple tokens, lost tokens, etc.).

The frame format of layer 2, based on IEC Standard 870-5-1, assures high data integrity. The data transmission services provided by layer 2 are

- Send-Data-With-Acknowledge
- Send-And-Request-Data-With-Reply
- Send-Data-With-No-Acknowledge
- Cyclic-Send-And-Request-Data-With-Reply

Layer 2 services are through service access points (SAPs) to the lower layer interface (LLI) and Fieldbus message specification (FMS) defined in layer 7 (DIN 19245 Part 2).

The main characteristics of layer 7 are

- Object-oriented client-server
- Modular structure with LLI and FMS
- Efficient messaging (index-short address, name-optional)
- Network management
- Connection and connectionless transmission

The Profitbus communication services are

- Context management (establish and release of connections)
- Variable access (cyclic and a-cyclic read and write)
- Domain management (download/upload memory)
- Program invocation management (start, stop, link programs)
- Event management (event handling)
- Virtual field device support (identification, status request)
- Object dictionary management (administration of OD)

Interface to the application process is from the application layer interface (ALI). However, LLI controls the data flow and mapping of the FMS services into layer 2. The user communicates with the application processes, using the following communication relationships:

Connection-oriented relationship is a logical peer to peer relationship between two application processes, master/master and master/slave. Data transfer among the nodes takes place in three phases:
1. Connection establishment phase, established prior data transmission with an "Initiate Service."
2. Data transfer phase, the actual data transmission between two communicating partners.
3. Connection release phase, released with the "Abort Service" when communication is no longer needed.

Connectionless communication relationship provides communications with several stations simultaneously in two forms:
1. Broadcast communication relationship, an unconfirmed FMS service, such as "Information Report," provided simultaneously to all master and slave nodes.
2. Multicast communication relationship, an unconfirmed FMS service simultaneously transmitted to a group of master and slave nodes.

Among typical functions of broadcast and multicast services are transmission of global alarm, synchronization, and other. These unconfirmed services are transmitted either with high or low priorities.

PROFITBUS OBJECT DICTIONARY (OD) AND ADDRESSING

ODs may be defined in any simple Profitbus-compatible device. The OD contains:

- Description
- Data type and structure
- Assignment between the device internal address of the object and the bus reference (Index/Name)

The structure of OD is depicted in Table 8.2.

TABLE 8.2
Structure of Profitbus OD

Header	Info about the structure of OD
Static list of types	A list of supported data types and structures
Static object dictionary	A list of static communication objects
Dynamic list of variable list	An actual list of known variable list
Dynamic list of programming invocations	A list of known programs

Static communication objects could be defined by the vendor or could be defined during system configuration. Profitbus recognizes the following static communication objects:

- Simple variable
- Array (simple variables of the same type)
- Record (simple variables of different types)

- Domain (data range)
- Event

Dynamic communication objects may be defined, predefined, or edited at any time during operation. The following dynamic objects are supported:

- ProgramInvocation
- VariableList (simple variables, arrays, records)

Addressing of Profitbus objects is defined as logical addressing in a form of a short address called **Index** (Unsigned 16). Also, for each device (object) an index is defined in the OD. In addition, optional addressing by name (symbolic name) or by physical address (a physical memory location) is supported. The Figure 8.4 illustrates a confirmed service using Index addressing.

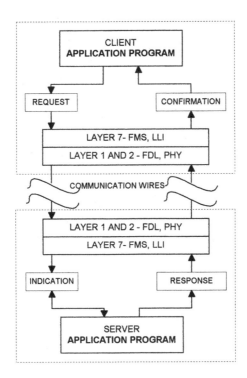

FIGURE 8.4   Confirmed service with Index addressing.

FMS APPLICATION SERVICES

The following application services are offered (services with * are supported by all Profitbus devices):

• Context management —	Initiate*
	Abort*
	Reject*
• OD management —	Get OD*
	Initiate Put OD
	Put OD
	Terminate Put OD

• Virtual field device support —	Status*
	Unsolicited Status
	Identify*
• Variable access —	Read
	Write
	Physical Read
	Physical Write
	Information Report
	Define Variable List
	Delete Variable List
• Program invocation management —	Create Program Invocation
	Delete Program Invocation
	Start, Stop, Resume, Reset, Kill
• Event management —	Event Notification
	Acknowledge Event Notification
	Alter Event Condition Monitoring
• Domain management —	Initiate Download Sequence
	Download Segment
	Terminate Download Sequence
	Initiate Upload Sequence
	Upload Segment
	Terminate Upload Sequence
	Request Domain Upload
	Request Domain Download

OTHER PROFITBUS FEATURES

The following is a list of additional Profitbus features:

Access protection: Optionally, each object may be protected against unauthorized access. Password and device groups may be defined in the OD.

Open and defined connections: *Open connections* are usually predefined by the vendor. It enables their communication with all other Profitbus stations without any need for additional configuration. *Defined connections* are fixed at the configuration time, and they can not be altered during the connection phases. It is to prevent unauthorized access during all communication phases.

Cyclic and acyclic data transfer: *Cyclic* data transfer is used, for example, for updates of remote inputs and outputs from a PLC. In Cyclic Data Transfer, only one variable is permanently read or written over the network. In an *acyclic* data transfer, the controller only occasionally accesses objects over the network.

Slave initiative: If the slave has the initiative attribute, it can send an unconfirmed FMS, such as an unsolicited alarm message, to the master.

Communication relationship list (CRL): CRL for simple nodes is predefined by the vendor. For more complex nodes, the CRL is loaded with the network management service locally or via the network. A communications reference list (CREF) consists of a static-predetermined part defined by the vendor and a dynamic part definable during startup for larger nodes.

Network management (FMA7): Enables vendor independent configuration, commissioning, and maintenance of the Profitbus, locally or remotely over the network. They are arranged in three groups:
1. Context management — establishes and releases connections.
2. Fault management — allows indication of faults, events, and reset of stations.

3. Configuration management — provides loading and reading of CRL, including access to variables, statistic counters, parameters of layers 1 and 2, identification of node components, and registration of stations.

Default management connection: Allows access to nodes for configuration and diagnostic devices.

Profitbus profiles: Application-specific definitions for communications for:
- Building automation
- Variable speed control
- Sensors and actuators
- PLCs
- Textile machines

Interoperability tests: Assure compliance with the defined profiles.

Internetwork: Connections to higher level networks are via Gateways.

Implementations: Profitbus can be implemented on any microprocessor that has asynchronous serial interface (UART). For example:

Compact implementation — the protocol and applications program resides on a single microprocessor; this is an advantage for simple slave devices.

Implementation with separate communication processor — used for master devices; due to separation of communication and application functions, all protocol and applications functions can be fully implemented.

Implementation with hardware support for layer 1 and 2 — for high performance, time critical applications, on separate chips or integrated in a standard microprocessor.

INTERBUS-S

GENERAL

Interbus-S was introduced as an open system for industrial controls on the sensor actuator level in 1987. It became a German standard DIN 19258. The Interbus-S protocol provides services to simple sensor/actuators as well as to more complex, so-called intelligent transducers. This lower level bus connects the field gear (sensor/actuators) to higher level programmable controllers or computers. It unites traditional voltage (i.e., 0 to 24 V), current (4 to 20 mA), and RS-232 connections into a single bus open system. Savings on design, wiring, and system startup characterize the Interbus-S connections. As a result, there are hundreds of thousands of field components on the market in compliance with Interbus-S open protocol. Automotive companies such as BMW, Mercedes, Audi, etc., as well as other industries, use Interbus-S for their automation networks.

Interbus-S-Club, providing interoperability tests and certifications for submitted components, assures compliance with the standard.

MAIN CHARACTERISTICS OF INTERBUS-S

Interbus-S is a hybrid protocol. Scan time on the sensor/actuator level is in the range of 1 to 10 ms. Messages at that speed are rather short, in the range of 16 to 32 bits. Such short messages are cyclical in nature. It also supports longer messages (i.e., several 16-bit words) transmitted at much greater time intervals — sequentially, or infrequently, as requested by the participating node. The protocol addresses all connected nodes on the network in the order that they are connected to the network (i.e., node 1, 2, etc.). Messages are sent out on the network simultaneously. Process data takes up 2 bytes, respectively, 4 bytes if the node parameters are included. Such data are transmitted in cyclical 16-bit frames. If acyclical data are requested, they are fit into the cyclical 16-bit frames sequentially. This imposes no greater load on the network traffic than transmission of 16 individual bits (i.e., 16 binary inputs or outputs) in a cyclical transmission. Inclusion of an acyclic transmission is by the circuitry of individual nodes. Due to the constant length of transmitted messages, their speed remains constant regardless of the data the frames contain (simple parameters or more complex messages).

INTERBUS-S TOPOLOGY

As seen in Figure 8.5, the Interbus-S topology is a ring with a master controlling the connected slaves. Duplex transmission and simple diagnostics characterize data transmission.

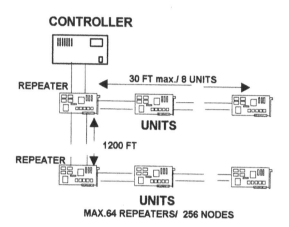

FIGURE 8.5 Interbus-S topology.

Transmission is standardized on a standard serial RS-485 communication over a pair of twisted shielded wires (or fiber optics for long transmission lengths). Signal repeaters may be used in the topology for repeating the signal as well as for configuration of the network and network diagnostics. The distance between individual repeaters can be up to 400 m (1200 ft). Each branch initiating from repeaters can support up to 8 nodes on a total length of 10 m (30 ft). The maximum number of repeaters is limited to 64, with a total of 256 nodes per master supported by the protocol. Due to the ring topology, the response (scan) time per master is constant and can be easily determined.

The above topology assures compatibility of data transmission with the higher level protocols in the hierarchy of industrial automation.

P-NET

GENERAL

P-Net was designed in the early 1980s to interface PLCs, applications-specific controllers, and intelligent sensors/actuators to control PCs, etc. on a single bus. Such installation then results in cost savings due to simpler design, installation, and cabling cost, as well as operating and maintenance cost. At the same time such installation provides more sophisticated online diagnostics, and offers possibilities for future expansions, while retaining the already installed field gear. P-Net became a standard in 1989 and is supported worldwide by a P-Net user organization.

BASIC COMMUNICATIONS AND CONFIGURATION

P-Net communicates via RS-485 serial communications, using a pair of twisted shielded wires. The maximum cable length is 1200 m, or 3600 ft without repeaters. The data transmission is asynchronous in NRZ coding. P-Net can handle up to 300 confirmed transactions per second from 300 independent addresses. Transmitted data are either process values (i.e., temperature, pressures,

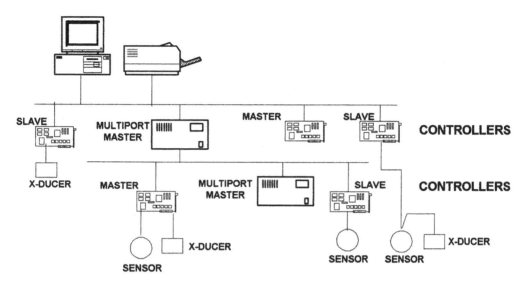

FIGURE 8.6   P-Net multinetwork architecture.

etc.) or as blocks of 32 independent binary values, such as start/stops, binary feedback from switch position, etc.

P-Net is a multiple master bus. The master sends requests to the addressed slave, which then returns an immediate response. When the master has finished its access, the token is passed to the next master. Passing the token from master to master is cyclical and time dependent.

P-Net offers multinet architecture, allowing inclusion of multiport masters in the P-Net (Figure 8.6). Multiport masters break down the entire network to more manageable subnetworks, corresponding to individual geographical or production areas. Each such area has its own master and slave nodes, interconnected via the multiport masters into a large P-Net. This makes the entire network more robust and less receptive to transmission errors from one subnet to another. The multinet structure allows direct addressing of the network device via the multiport masters in a particular slave.

P-Net does not allow different categories or conformance classes. All P-Net devices are alike in their communication interfaces, which minimizes design and configuration time. It also provides for interchangeability of P-Net devices during system startup and operation. Node configuration time is also greatly reduced due to minimal configuration of each node, which consists of an address setting and of additional settings equal to the number of masters in the network in the master node. Most P-Net devices are sealed and the settings can be done through the network. Parameters of P-Net devices can be accessed via the network using a so-called "Softwire" list, which is automatically generated while the application program is compiled. Real-time network traffic is ensured by restricting the length of each frame to 56 bytes. P-Net devices divide longer messages into consecutive frames for transmission. P-Net slave devices can handle more then just I/O functions. They can provide full PID control functions as well as calculations. Master nodes usually contain scaling of values, set-point variables, clock functions, etc. Slaves also provide error-checking functions and report errors to the master node when the slave is poled.

P-Net Channel Structure

The collection of variables related to a process object is called a **channel**. Channels contain a collection of data necessary to support a certain control function for the process object, as well as network management (error checking) and maintenance functions. Each channel, designated by channel type, has 16 registers with their assigned Softwire logical address, description, and associated values. Channels may have different designations, such as PID channel, communications channel, printer channel, and service channel. The service channel must be included in all nodes, whether they represent simple I/O structures or a collection of several different channels. The service channel contains information about the node address, serial number, manufacturer, error data, and other data associated with that particular node.

ACCESSING P-NET

A PC can be a P-Net master, accessing the P-Net via a Fieldbus management system called VIGO. Real-time exchange of data between Windows applications and physical objects is via OLE2 automation product (object linking and embedding). This enables the use of standard programs, such as Visual Basic, Visual C++, spreadsheets, databases, etc., with P-Net networks.

P-Net can use the same microprocessor that controls the main task of the node as well as the communication. That makes P-Net implementation into field devices by various vendors cost-effective.

CAN: CONTROL AREA NETWORK

GENERAL

Originally, CAN was developed by BOSCH for use in automobiles. Later on, CAN became ISO Standard 11898 for high speed and 11519-1 for low speed communications. Today, there are both individual hardware components and microcomputers using CAN in automobiles and in industrial applications.

CAN is used in automobiles at communications speeds of 200 kbps to 1 Mbps to control such functions as:

- Engine control (ignition, fuel injection)
- Interior control, such as control of lighting, air-conditioning, central door lock, adjustments of mirrors, etc.
- Transmission control, including antiblock-system, acceleration skid control, etc.
- Control of communication systems, such as telephones, radios, etc.
- System diagnostics control

In industrial applications, CAN is used primarily for its low-cost, relatively fast startup time, expandability, and for its good resistance to electromagnetic disturbances. Thanks to universal modules, CAN is used in various industrial applications. The most attractive applications for CAN are in areas of high-speed communications.

COMMUNICATION

CAN communicates on a two-wire system CAN_H and CAN_L, with end of line (EOL) resistors @ 120 Ω The generator has two states — active (called dominant), and passive (called recessive). In recessive state (logical 1), voltage levels on both lines are about equal (i.e., 2.5 V). In dominant state (logical 0), the voltage level on CAN_H is, lets say, 3.5 V, while on CAN_L it is only 1.5 V. The receiver uses a method called wired AND, which allows changing logical level 1 (recessive) by another receiver to logical level 0 (dominant). This allows acknowledgment of received messages and provides for error checking.

High-speed CAN communicates at a speed of 125 kbps to 1 Mbps on a maximum trunk distance of 40 m or 120 ft to 2 to 30 connected nodes. The transmission speed is reduced with the increased length of the communications trunk.

There are four types of messages defined in CAN:

1. Data frame
2. Remote frame
3. Error frame
4. Overload frame

Messages are sent from node to node or are broadcasted to all nodes across the network. The receiving nodes acknowledge receipt of the message with a dominant bit in the ACK field. Error checking is provided in the error frame of the message format. The CAN protocol provides high-speed, high-reliability communication with a very low error rate. As such, it became a preferred choice for communication on the lowest, sensor–actuator levels of systems hierarchy.

FIP: FIELD INSTRUMENTATION PROTOCOL

FIP was developed by a group of French users, manufacturers, and researchers for interoperability of field devices. FIP became a French standard in the late 1980s. It uses three layers of the OSI model, the physical, data link, and application layers.

The standard can be used for networking of simple sensors and actuators as well as intelligent field devices. Furthermore, it can be used for interconnection of lower level controllers. From the hierarchical point of view, it represents a true fieldbus.

The **physical layer** of the FIP protocol supports transmission over a twisted shielded pair of wires as well as over fiber optic cables. The number of nodes per driver without use of a repeater is 60, with a maximum cable length of 3000 ft at a maximum transmission rate of 2.5 Mbps. The number of nodes or the distance limitations can be exceeded if the electrical characteristics of the network, such as attenuation, impedance, and propagation times are kept below their maximum values. The transmission mode specified for FIP is synchronous with Manchester data encoding. Medium access control (MAC) is centralized.

The **data link layer** provides transmission services. It also assigns identifiers for *periodic messages* and queues them for transmission. Multiple identifiers as well as queues are assigned to messages that require different scan times. Pending *messages on request* are assigned to queues in the data link layers of the sending and receiving nodes. The queued messages are then scheduled, as are the periodic messages, and sent across the network. Application services provide local read/write services for periodic messages, and remote read/write services for messages on demand. The services are grouped into classes, such as sensor, actuator, I/O concentrator, programmable language controller, operator, and programming consoles.

The greatest strength of FIP is in its guaranteed response time and number of messages transmitted. This applies, especially, to transmitting periodical messages. Transmission of messages on request is less efficient and its interpretation can be confusing for the user, due to lack of time stamps associated with the messages.

ASI: ACTUATOR SENSOR INTERFACE PROTOCOL

ASI was developed to interface devices on the lowest level on the system architecture, on the sensor/actuator level. Furthermore, the protocol is restricted to transmission of binary values. Its purpose is to provide savings to the end users by reducing wiring costs, replacing expensive star-like wiring (from controllers to sensors and actuators) with an inexpensive, low-cost interface protocol.

The ASI protocol either can be implemented on a single chip of an intelligent sensor and/or actuator or it can be on a controller providing network interface to the connected field devices. Such an arrangement provides interface to sensors and actuators manufactured by different manufacturers connected to the same network. The connection can be via inexpensive two-wire unshielded cables in a network topology (bus, star, ring, etc.) most suitable for the given application. Cable coupling and assembly is via modular connectors, making the ASI cabling inexpensive for installation. The signal and 24 VDC supply voltage are provided on the same cable. Due to the restriction of data transfer to binary signals (3 to 4 bits), the transmission time achieved is a relatively short 5 ms.

In system architecture, the ASI master (one master per network) connects the ASI network to a higher level controller, such as a PC, PLC, microcomputer, etc. The ASI slaves (31 slaves per network) are then connected to the ASI bus (cable), which also carries the 24 VDC power source. Serial transfer of information is bidirectional between the controller and the connected field devices, which are polled sequentially. There can be a maximum of 124 binary sensors and actuators connected to one network.

Each slave is assigned a permanent address, which is stored in the ASI master. The master records the slave IDs, and also records changes of IDs for the replaced slaves. Besides polling, the master also provides network management functions, such as network initialization, slave diagnostics, on-demand download of parameters to the slaves, error reports to the higher level controller, and other functions.

ASI is a simple, inexpensive, and reliable solution to provide manufacturing facilities with communications on the lowest level of hierarchy with high concentration of binary field points.

HART: HIGHWAY ADDRESSABLE REMOTE TRANSDUCER PROTOCOL

Distributed processing allows designers and users of control systems to obtain and process information on the lowest layers of systems hierarchy, on the (unitary) controller and sensor/actuator/transducer levels. Sensor manufacturers have worked diligently on development of so-called intelligent sensors to provide the following features:

- Networking of sensors/transducers in a multidrop fashion. This results in savings on design and installation cost due to eliminating wiring up of individual sensors in a star-like configuration to their respective controllers. Such sensors/transducers contain communications protocol and are addressable individually on the network.
- Access of sensor/transducer parameters over the network. While analog sensors/transducers (i.e., 4 to 20 mA, 0 to 20 V) can provide only one information to their respective controllers — the present value (PV), intelligent sensor/transducers provide a variety of data and setup parameters. For example, an intelligent sensor/transducer may contain, besides the present value, also information on setup and tuning parameters, PID algorithms, and other parameters. Since they are accessible via the network, such sensors can be set up and calibrated either from the controller via a hand-held device or from the control room via a networked OWS. This results in a shorter setup, calibration, and validation time, and ease of troubleshooting during failure or regular maintenance (calibration) of the field gear. Since most of such sensor/actuators are located in inaccessible places, for example, on distribution piping at high elevations inside a power plant, their maintenance and calibration could be a very cumbersome task requiring allocation of excessive resources and manpower. Thus, intelligent sensor/transducers save not only on initial costs, but also on operating and maintenance costs.
- Network communication. Intelligent sensor/transducers can communicate over their respective networks, to other nodes (i.e., controllers) connected to the network. Due to their low cost, they comply with a limited but essential number of networking features and network management functions such as addressing, directory services, fault analysis, etc.

HART is an open protocol for distributed sensor applications. It was developed by Rosemount in the mid 1980s, and became an open protocol in the late 1980s. Today, further development of the protocol and its support is administered by a nonprofit Hart Communications Foundation, which also owns the trademarks and protocol rights. There is also a HART user group formed in the early 1990s to support the protocol. Members of the user groups are primarily instrument manufacturers interested in standard communications at the lowest/sensor/actuator layer of systems hierarchy.

HART protocol enables two-way communications to smart (digital) transducers on the same network with analog 4 to 20 mA devices. This feature provides means of networking existing analog devices on the same network with smart digital sensor/transducers, thus preserving the owner's investment in previously installed field devices. This is achieved by superimposing frequency shift keying (FSK) digital signals on the low-level, 4- to 20-mA analog signals. Since FSK is phase continuous, there is no interference with the analog signal. The protocol provides for two-dimensional error checking. The "logical 1" is transmitted at 2200 Hz, while the "logical 0" is at a frequency of 1200 Hz. The protocol guarantees two to three updates per second in a "Request–Response" mode, and optional three to four updates per second in a so-called "burst" mode.

The network topology can be point-to-point, in a star-like configuration, for analog, and combined analog and digital transmission or multidrop topology for digital only transmission. The cable length, depending on the cable and configuration, can be up to 10,000 ft (3048 m).

The HART protocol communication stack implements the 7th (applications), 2nd (data link), and 1st (physical) layers of the ISO/OSI reference model.

The application layer provides formatted data to the application programs of compatible devices manufactured by different manufacturers. The HART commands are organized into universal, common-practice and device-specific commands.

Universal commands are part of **all** HART-compatible devices. They provide information, such as:
  - Read — manufacturer, model, serial number, device type, device status, PC, engineering units, range values, up to 4 predefined dynamic variables, etc.
  - Read/Write — tag number (8 character), description (16 character), date, message (32 character), final assembly number, etc.
  - Write — polling address.

Common-practice commands are to access maintenance-related functions and information:
  - Read — up to 4 dynamic variables, etc.
  - Read/Write — dynamic variable assignments, etc.
  - Write — sensor serial number, time constant, transmitter range, set zero, set span, set fix output current, trim PV zero, write PV units, trim zero and gain, write square root or linear transfer functions, perform master reset, perform self-test, etc.

Device-specific commands access unique features of particular field devices set up during configuration and startup.
  - Model-specific functions, start/stop, clear totalizer, select primary variable, calibration options, PID loop tuning, etc.

Each message reply includes information on communications error (if one occurred) and state of the slave device (device malfunction, configuration change, etc.).

HART has recognized the need of individual users (i.e. sensor manufacturers) to include specific device, application, or functionality-related information without changing the basic structure of the protocol. For that purpose it includes so-called classes and subclasses. Such a subclass may be a profile of an instrument, for example, for pressure, temperature, or flow measurement. Manufacturers may write their device-related specific information in a device description language (DDL), compile it with a DD compiler, and make it available to the users over the HART network.

The HART protocol user-friendly features are the cause of wide acceptance of the protocol by sensor manufacturers and end users alike. Protection of the investment in instrumentation, savings on wiring costs, and provision of remote access over the network to set up and calibrate made the use of the protocol an attractive choice for smart sensors.

CONCLUSION

Like all communications-related fields, protocols associated with lower level industrial networks are in constant evolution. There are new drivers and protocols offered by manufacturers of instrumentation, engineering associations, manufacturers of digital control systems (DCS), etc., at an unprecedented pace. As with everything else, it is up to the smart consumer to pick the protocol(s) and/or drivers most suitable for the given application or, better yet, for the given plant. Considerations for their selection should include (besides their reliable interoperability) commonsense items, such as protection of the owner's investment in previously installed systems, long-term support of the protocol, and reduction of the risk associated with obsolescence of the protocol or driver.

Although there are efforts associated with standardization of communication protocols on every system level, experience shows that the best protocols or drivers are the ones that have matured in popularity over time. They consequently become de facto industrial standards by "popular demand," rather than by committee effort. It is more prudent for owners to select from de facto industry standard protocols or drivers already on the market rather than wait for "yet-to-be-developed" standard protocols and drivers. This, however, does not mean that one should not support the effort of the industry toward standardization. In the meantime, end users need more information related to evaluation and selection of protocols and drivers available on the market. The same applies for compliance evaluation of individual products with their associated protocols and drivers.

9 Canadian Automated Building (CAB) Protocol

Viktor Boed

CONTENTS

GENERAL

In effort to standardize communications between different vendors' building automation systems (BAS), Public Works Canada (PWC) has developed and published in 1992 a standard known as Canadian Automated Building (CAB) protocol. The standard facilitates communications among proprietary BAS utilized for environmental control (temperature, humidity, and air quality), energy management, fire alarm and life safety monitoring, security control and monitoring, and lighting control in governmental buildings.

The CAB standard was published by PWC and offered as public domain. To make proprietary BAS interoperable with each other on a common local area network (LAN), systems integrators, building owners, and controls and automation systems vendors can utilize the CAB standard.

Public Works Canada welcomes inquires or expressions of interest from other governments, agencies or private sector organizations respecting the use of the CAB Protocol for their applications...

is an invitation for the controls and automation industry to develop protocols which would comply with the CAB standard. It is also an invitation for building owners, developers, and engineers to use the above standard in design and in the competitive procurement of BAS.

CAB TOPOLOGY

In the absence of a common communications protocol, individual controls and automation systems communicate with their controllers via their own proprietary protocols. As shown on Figure 9.1, vendor-specific controllers communicate with their unitary and application-specific controllers on "vendor-specific" networks. Vendor-specific controllers on a CAB network can share information on a backbone LAN — (**Public Works Canada PWC LAN**), either via so-called **CAB-Gateways** or directly, if the controller or PC OWS is CAB compatible.

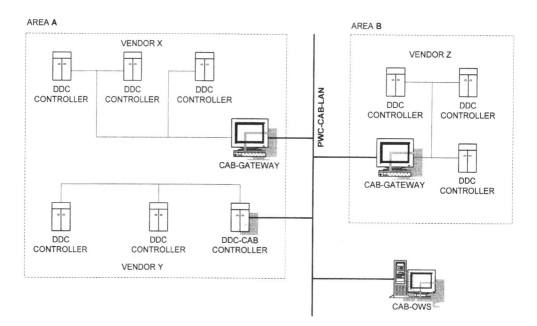

FIGURE 9.1   CAB PWC LAN topology.

Vendor-specific controllers with CAB protocols or CAB-Gateways are so called "network node devices" residing within their respective "areas." Areas are interconnected via the backbone or via

bridge/interface devices to the LAN or wide area network (WAN). Each network node device can be accessed and/or modified, utilizing CAB protocol services.

The CAB protocol shares data among individual controls and automation systems and provides local and distributed monitoring of controls and automation systems over the network. It also allows data exchange with other computer-based, real-time and/or administrative systems connected to PWC CAB LAN.

THE CAB PROTOCOL MODEL

CAB protocol utilizes the seven layer architecture of the ISO Model. However, the CAB standard defines the "application" layer by **CAB point types**, **application services**, and **other CAB-specific items**.

The application layer is designed to support communications among application programs residing in different vendor's controllers or OWS' over the network. This is provided via so-called **confirmed** (**request–acknowledge–process–reply–acknowledge**), **unconfirmed** (**request–acknowledge**), and **broadcast** services (see Figures 9.2, 9.3, and 9.4).

Protocol transactions, requests, and replies are encoded into data elements called protocol data units (PDUs). Each data unit is self-contained, independent, in a form of a "datagram" transmitted as part of "connectionless" communications mode.

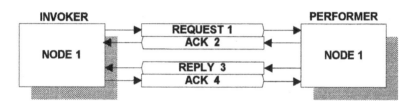

FIGURE 9.2   Confirmed service transaction.

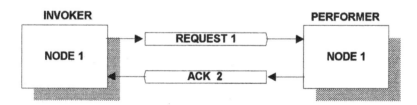

FIGURE 9.3   Unconfirmed service transaction.

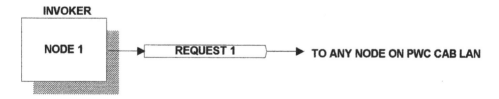

FIGURE 9.4   Broadcast service transaction.

CAB POINT TYPES

A CAB point is a representation of a physical point or point defined in the program (pseudo-point) residing in CAB-compatable controllers or CAB gateways. A CAB point is composed of name and point-specific attributes.

The following point types are defined as CAB points: **Digital input (DI)**, **Supervised digital input (SDI)**, **Digital output (DO)**, **Analog input (AI)**, **Analog output (AO)**, **Controller (CL)**, **Tri-State input (TI)**, and **Tri-State output (TO)**.

The following tables provide an overview of CAB defined point types. The information is tabulated to provide a quick overview for facilities engineers and managers who are not communications experts. Readers interested in detailed CAB protocol specification should contact PWC for the full CAB documentation.

The following symbols are used in the tables:

* *Simple change of state/change of value (COS/COV)* — any change will cause COS/COV notification
$ *Standard COS/COV* — changes greater or less than the present value ± differential will cause COS/COV notification
# *Attribute eligible for trending*
@ *Default attribute* — default attributes returned under certain conditions as response to CAB service request (bold type in the tables)

DIGITAL INPUT

DI — a two-state physical or pseudo input.

TABLE 9.1
DI — Two-State Input

Attributes	Response	Symbols	Read/Write	Description/Options
Present Value	**"PV On Text"**	*	R	PV depends on contact condition
	"PV Off Text"	#		
PV On Text	"Open"	*	R/W	On, Run, Up, etc.
PV Off Text	"Closed"	*	R/W	Off, Stop, Down, etc
Contact cond.	"On"	*	R	On or Off
State	**"Enable"**	*	R/W	**Ena** = PV used in control logic
				Dis = last PV used in control logic
Override	"Off"	*	R/W	Operator On/Off
Alarm Status	**"Normal"**	*#	R	Norm/OnAlm/OffAlm
				"Alm" when "Cond.to Alm" = On or Off
				With/matching "Cont.Cond."
				Paired DI/DO have time delay
				0–180 before ALM
Condition to Alarm	"On"	*	R/W	**On/Off** = cont.cond.for Alarm
				None = deactivates Alm reporting
Alarm category	"Critical"	*	R/W	Cautionary/Maintenance
Runtime Ena	"On"	*	R/W	Activates totalizer
Runtime value	"123.4"	*	R/W	Hours; 0 = reset
Runtime time	"12:34"	*	R	Last time reset

Attributes	Response	Symbols	Read/Write	Description/Options
Runtime date	"09-08-95"	*	R	Last reset date
Runtime limit	"18000"	*	R/W	In minutes to Alarm
Runtime Alm. Cat.	"Maintenance"	*	R/W	Same as Alm Categories
Runtime reset value	"200"	*	R/W	Upper limit of Rt. value in hours
COS count on	"True"	*	R/W	Counter Ena/Disable
COS count	"234"	*	R/W	Total COS count
COS count limit	"1000"	*	R/W	Max. COS count = Alarm
COS count window	"1"	*	R/W	No. of COS counts in "x" min (sliding)
COS count Alm. category	"Maintenance"	*	R/W	Critical/Cautionary/Maintenance
Pt.Alm.Text 1	"Send HVAC tech"	*	R/W	Up to 80 characters
Pt.Alm.Text 2	"Check Temp"	*	R/W	Up to 80 characters
Point Text 1	**"HW pump 1"**	*	R/W	Up to 40 characters pt. name expansion
Point Text 2	**"South loop"**	*	R/W	Same as above

SUPERVISED DIGITAL INPUT

SDI — a group of two inputs: One = status of an actual contact (reported to "PV," "Contact Condition"); Second = supervisory circuit monitoring contact wiring (reported as NORMAL/TROUBLE to "Supervisory Status" — Digital or analog circuit).

TABLE 9.2
SDI — Group Inputs

Attributes	Response	Symbols	Read/Write	Description/Options
Present Value	**"PV On Text"**	*	R	PV depends on contact condition
	"PV Off Text"	#		
PV On Text	"Open"	*	R/W	On, Run, Up, etc.
PV Off Text	"Closed"	*	R/W	Off, Stop, Down, etc
Contact Cond.	"On"	*	R	On or Off
State	**"Enable"**	*	R/W	**Ena** = PV used in control logic **Dis** = last PV used in control logic
Override	"Off"	*	R/W	Operator On/Off
Alarm status	**"Normal"**	*#	R	Norm/SecureAlm/TroubleAlm **"SecureAlm"** when "Cond.to Alm." = On/Off w/matching "Cont.cond." **"TroubleAlm"** when "Supervisory.Stat." = "Trouble"

Attributes	Response	Symbols	Read/Write	Description/Options
Cond.to alarm	"On"	*	R/W	On/Off when"Cont.cond" = SecureAlm condition
Alarm mode	**"Secure"**	*#	R/W	**ACCESS** – Alm not reported **SECURE** – when "Cont.Cond." = "Cond.toAlm" = ALM
Schedule (sets Alm mode)	"16"	*	R/W	16 daily schedules @ up to 100 time/ value events; empty = disabled
Calendar	"01-02-95"	*	R/W	Which schedule should be active @ day
Supervisory status	"Normal"	*#	R	**TROUBLE** – wiring problem = CRTALM in "AlmMode"
COS count on	"True"	*	R/W	Counter Ena/Disable
COS count	"234"	*	R/W	Total COS count
COS count limit	"1000"	*	R/W	Max. COS count = Alarm
COS count window	"1"	*	R/W	No. of COS counts in "x" min (sliding)
COS Count Alm Category	"Maintenance"	*	R/W	Critical/Cautionary/Maintenance
Pt.Alm.Text 1	"Send fire tech"	*	R/W	Up to 80 characters
Pt.Alm.Text 2	"Check Zone 1"	*	R/W	Up to 80 characters
Point Text 1	**"Fire Alarm"**	*	R/W	Up to 40 char. pt. name expansion
Point Text 2	**"Zone 1 loop"**	*	R/W	Same as above

DIGITAL OUTPUT

DO — a two-state physical or pseudo-output.

TABLE 9.3
DO — Two-State Digital Output

Attributes	Response	Symbols	Read/Write	Description/Options
Present value	**"PV On Text"** **"PV Off Text"**	* #	R	PV depends on contact condition
PV On Text	**"Open"**	*	R/W	On, Run, Up, etc.
PV Off Text	**"Closed"**	*	R/W	Off, Stop, Down, etc
Contact cond.	"On"	*	R	On or Off
State	**"Enable"**	*	R/W	**Ena** = PV used in control logic **Dis** = last PV used in control logic
Operating mode	**"Auto"**	*	R/W	**Manual** — op. override
Feedback value	"On"	*	R	Actual status On/Off

Attributes	Response	Symbols	Read/Write	Description/Options
Alarm status	**"Normal"**	*#	R	Norm/OnAlm/OffAlm/ NtEqToFeedbackAlm "Alm"-Cond. to Alm.= On/Off w/matching "Contact Condition"
Cond.to Alarm	"On"	*	R/W	On/Off – cont.cond.for Alarm None – deactivates Alm reporting
Alarm category	"Critical"	*	R/W	Cautionary/Maintenance
Schedule (sets Alm mode)	"16"	*	R/W	16 daily schedules @ up to 100 time/ value events; empty=disabled
Calendar	"01-02-95"	*	R/W	Which schedule should be active @ day
Runtime Ena	"On"	*	R/W	Activates totalizer
Runtime value	"123.4"	*	R/W	Hours 0=reset
Runtime time	"12:34"	*	R	Last time reset
Runtime date	"09-08-95"	*	R	Last reset date
Runtime limit	"18000"	*	R/W	In minutes to Alarm
Runtime Alm Cat.	"Maint."	*	R/W	Same as Alm Categories
Runtime reset value	"200"	*	R/W	Upper limit of Rt.value in hours
COS count on	"True"	*	R/W	Counter Ena/Disable
COS count	"234"	*	R/W	Total COS count
COS count limit	"1000"	*	R/W	Max. COS count = Alarm
COS count window	"1"	*	R/W	No. of COS counts in "x" min (sliding)
COS count Alm category	"Maintenance"	*	R/W	Critical/Cautionary/Maintenance
Pt.Alm.Text 1	"Send HVAC tech"	*	R/W	Up to 80 characters
Pt.Alm.Text 2	"Check HW loop"	*	R/W	Up to 80 characters
Point Text 1	**"HW Valve1"**	*	R/W	Up to 40 char.pt.name expansion
Point Text 2	**"South loop"**	*	R/W	Same as above

ANALOG INPUT

AI — a continuous physical or pseudo-input.

TABLE 9.4
Analog Input

Attributes	Response	Symbols	Read/Write	Description/Options
Present value	**"72"**	$#	R	

Attributes	Response	Symbols	Read/Write	Description/Options
PV eng.unit	**"DEGF"**	*	R/W	DEGC, %pos, psi
State	**"ENA"**	*	R/W	Enable/Disable
Override	"Off"	*	R/W	On=Op.override to PV
Reliability	"Good"	*#	R	Bad=CritAlm
PV COV differential	"1"	*	R/W	PV ± deviation
Range 0 Scale	"50"	*	R/W	LowestPV=CrAlm & Bad
Range full scale	"120"	*	R/W	HighestPV=CrAlm & Bad
High-2 Alm.Lim.	"90"	*	R/W	HiHiAlm limit
High-1 Alm.Lim.	"80"	*	R/W	HiAlm limit
Low-1 Alm.Lim.	"68"	*	R/W	LoAlm limit
Low-2 Alm.Lim.	"60"	*	R/W	LoLoAlm limit
Alarm deadband	"2"	*	R/W	Alm Limit ± Dev.
Alarm Status	"Normal"	*#	R	(NORMAL/HI1/HI2/LO1/ LO2/Range0/RangeFULL/ RelBAD)
Point Alarm Text1	"Send HVAC tech"	*	R/W	Up to 80 Characters
Point Alarm Text2	"24 hr response"	*	R/W	Up to 80 Characters
Point Text 1	**"Director's Office"**	*	R/W	Up to 40 Characters
Point Text 2	**"Likes it hot"**	*	R/W	Up to 40 Characters

ANALOG OUTPUT

AO — a continuous physical or pseudo-output, including PID attributes.

TABLE 9.5
AO — Analog Output

Attributes	Response	Symbols	Read/Write	Description/Options
Present value	**"50%"**	$#	R/W	(0–100%)Op.entry in MANUAL "Op.Mode"
State	**"Enable"**	*	R/W	Operator command
Operating mode	**"Auto"**	*	R/W	**Manual** – PV override by operator
Control mode	**"Logic"**	*	R/W	PV set by control logic **Operator** = set by Operator
Measured value	"71.5"	$#	R	=AI PV
MV eng. units	**"DEGF"**	*	**R/W**	**DEG C, psi, etc.**

Attributes	Response	Symbols	Read/Write	Description/Options
Reset value	"72"	$#	R	PV of AI to reset SPV
RV eng. units	"DEGF"	*	R/W	DEG C, psi, etc.
Setpt value	**"72"**	$#	R/W	SPV of the PID loop. Can be set by the operator
AO feedback Value	"49.5%"	$#	R	Feedback
Prop.value	"10"	*	R/W	Proportional constant
Integral value	"5"	*	R/W	Integral constant
Derivative value	"1"	*	R/W	Derivative constant
Deadband value	".5"	*	R/W	For the PID control loop
Bias value	".2"	*	R/W	Added to SPV in **Auto** or **Logic** modes
Deviation AlmHigh	"2"	*	R/W	Max. dev. of MV above SPV
Deviation AlmLow	"2"	*	R/W	Max. dev. of MV below SPV
Alarm differential	"2"	*	R/W	Deadband=SPV ± DevAlm/ ± AlmDif
Alarm status	**"Normal"**	*#	R	(Normal/DevHIGH/DevLOW)
PV COV differential	"2"	*	R/W	± Dev. from PV/AOfdbck
MV COV differential	"1"	*	R/W	± Dev. from MV/RSV/SPV
Scheduled attribute	"SPV"	*	R/W	PV/SPV will trigger "Schedule"
Schedule	"1"	*	R/W	16 daily schedules @ up to 100 time/ value events; empty=disabled
Calendar	"01-02-95"	*	R/W	Which schedule will be active @ day
Point Alarm Text 1	"Send HVAC tech"	*	R/W	Up to 80 Characters
Point Alarm Text 2	"24 hr response"	*	R/W	Up to 80 Characters
Point Text 1	**"Main Valve"**	*	R/W	Up to 40 Characters
Point Text 2	**"East Zone"**	*	R/W	Up to 40 Characters

CONTROLLER

CL — virtual point associated with the control program in the controller.

TABLE 9.6
Controller

Attributes	Response	Symbols	Read/Write	Description/Options
Present value	**"50%"**	$#	R/W	(0–100%)Operator entry in **manual** mode

Attributes	Response	Symbols	Read/Write	Description/Options
State	**"Enable"**	*	R/W	Operator command
Operating mode	**"Auto"**	*	R/W	**Manual** — PV override by operator
Control mode	**"Logic"**	*	R/W	PV set by control logic **Operator** — set by operator
Measured variable	"R123T"	*	R/W	Pt ID
Measured value	"71.5"	$#	R	=AI PV
MV eng. units	**"DEGF"**	*	**R/W**	**DEG C, psi, etc.**
Reset variable	"AH1FDT"	*	R/W	Resets SPV
Reset value	"60"	$#	R	PV of RSV to reset SPV
RV Eng. Units	"DEGF"	*	R/W	DEG C, psi, etc.
Setpt Value	**"72"**	$#	R/W	SPV of the PID loop.Can be set by the operator
AO fdback variable	"VLVFB"	*	R/W	FB from the valve
AO feedback value	"49.5%"	$#	R	Feedback
Prop.value	"10"	*	R/W	Proportional constant
Integral value	"5"	*	R/W	Integral constant
Derivative value	"1"	*	R/W	Derivative constant
Deadband	".5"	*	R/W	For the PID control loop
Bias	".2"	*	R/W	Added to SPV in **Auto** or **Logic** modes
Deviation AlmHigh	"2"	*	R/W	Max. dev. of MV above SPV
Deviation AlmLow	"2"	*	R/W	Max. Dev. of MV below SPV
Alarm diff.	"2"	*	R/W	Deadband=SPV ± DevAlm/ ± AlmDif
Alarm status	**"Normal"**	*#	R	(Normal/DevHIGH/DevLOW)
PV COV differential	"2"	*	R/W	± Dev. from PV/AOfdbck
MV COV differential	"1"	*	R/W	± Dev. from MV/ReSV/STPV
Schedule (Setpoint)	"1"	*	R/W	16 daily schedules @ up to 100 time/ value events; empty=disabled
Calendar	"01-02-95"	*	R/W	Which schedule will be active @ day
Point Alarm Text1	"Send HVAC tech"	*	R/W	Up to 80 Characters
Point Alarm Text2	"24 hr response"	*	R/W	Up to 80 Characters
Point Text1	**"Main Valve"**	*	R/W	Up to 40 Characters
Point Text2	**"East Zone"**	*	R/W	Up to 40 Characters

TRI-STATE INPUT

TI — associated with actual or virtual field point.

TABLE 9.7
TI — Tri-State Input

Attributes	Response	Symbols	Read/Write	Description/Options
Present value	**"Off"**	*#	R	PV1,2,or 3 text
PV 1 Text	"Off"	*	R/W	(Off, Closed, etc.)
PV 2 Text	"Slow"	*	R/W	(Position 1, etc.)
PV 3 Text	"Fast"	*	R/W	(Position 2, etc.)
Contact cond.	"Off"	*#	R	Of PV1 value
State	**"Enable"**	*	R/W	**Disable** — PV no change
Override	"Off"	*	R/W	**On** — Operator override
Alarm status	**"Normal"**	*	R	(Normal/PV1Alm/PV2Alm/PV3Alm)
Cond. to Alm reporting	"None"	*#	R/W	PV1/PV2/PV3/None — disable Alm
Alm. category	"Maintenance"	*	R/W	(Critical, cautionary maintenance)
Runtime Ena	"On"	*	R/W	(On/Off)
Rt off state	"Off"	*	R/W	PV1,PV2=On
Rt value	"123"	*	R/W	Hr to 1 dec. place 0=Rt value reset
Rt time	"12:34:"	*	R	Last reset time
Rt date	"12-09-95"	*	R	Last reset date
Rt limit	"60000"	*	R/W	Max. value in min
Rt Alm. Cat.	"Maintenance"	*	R/W	(Crit./caut./maint.)
Rt reset value	"60000"	*	R/W	Max.Rt.value — resets Rt. value, time, date
COS count on	"On"	*	R/W	COS counter On/Off
COS count	"222"	*	R/W	Reset on min.interval
COS ct.lim.	"500"	*	R/W	Max. count=Alm
COS ct. window	"60"	*	R/W	No. of cts in the past 60 min
COS ct.Alm.Cat	"Maint"	*	R/W	(Crit./caut./maint.)
Pt Alm. Text 1	"Send EQ tech"	*	R/W	Max. 80 char.
Pt Alm. Text 2	"Call chem. lab"	*	R/W	Max. 80 char.
Point Text 1	**"Fan Off"**	*	R/W	Max. 40 char.
Point Text 2	**"2speed ex.fan"**	*	R/W	Max. 40 char.

TRI-STATE OUTPUT

TO — associated with actual or virtual field point.

TABLE 9.8
TO — Tri-State Output

Attributes	Response	Symbols	Read/Write	Description/Options
Present value	**"Slow"**	*#	R	PV1, 2, or 3 text
PV 1 Text	**"Off"**	*	R/W	(Off, Closed, etc.)
PV 2 Text	**"Slow"**	*	R/W	(Position 1, etc.)
PV 3 Text	**"Fast"**	*	R/W	(Position 2, etc.)
Contact Cond.	"Closed"	*#	R	of PV1 value
State	**"Enable"**	*	R/W	**Disable** — PV no change
Operating mode	"Auto"	*	R/W	**Manual** — Operator override
Feedback value	"Closed"	*#	R	PV1/PV2/PV3 TI state
Alarm status	**"Normal"**	*	R	(Normal/PV1Alm/ PV2Alm/PV3Alm/Not EqToFdbckAlm)
Cond.to Alm.	"None"	*#	R/W	None — disable Alm reporting; (PV1/2/3/NotEqToFdbAlm/None)
Alm. category	"Maintenance"	*	R/W	(Critical, cautionary maintenance)
Schedule	"16"	*	R/W	16 Schedules/24 hr @upto 100 time/ value events
Calendar		*	R/W	no schedule
Runtime Ena	"On"	*	R/W	(On/Off)
Rt off state	"Off"	*	R/W	PV1,PV2=On
Rt value	"123"	*	R/W	Hr to 1 dec. place 0=Rt value reset
Rt time	"12:34"	*	R	Last reset time
Rt date	"12-09-95"	*	R	Last reset date
RT limit	"60000"	*	R/W	Max. value in min.
Rt Alm. Cat.	"Maintenance"	*	R/W	(Crit./caut./maint.)
Rt reset value	"60000"	*	R/W	Max.Rt.value — resets Rt. value, time, date
COS count on	"On"	*	R/W	COS counter On/Off
COS count	"222"	*	R/W	Reset on min. interval

Attributes	Response	Symbols	Read/Write	Description/Options
COS ct. lim.	"500"	*	R/W	Max. count=Alm
COS ct. window	"60"	*	R/W	No. of cts in the past 60 min
COS ct.Alm.Cat	"Maint."	*	R/W	(Crit./caut./maint.)
Pt Alm Text 1	"Send EQ tech"	*	R/W	Max. 80 char.
Pt Alm Text 2	"Call chem. lab"	*	R/W	Max. 80 char.
Point Text 1	**"Fan slow speed"**	*	R/W	Max. 40 char.
Point Text 2	**"2speed ex.fan"**	*	R/W	Max. 40 char.

CAB PROTOCOL SERVICES

Requests to remote CAB controllers or OWS on a CAB network are defined by CAB-Service-Argument. A CAB-Confirmed service responds either with a Complex or Simple Service Result. CAB-Unconfirmed-Service and CAB-Broadcast-Service receive no confirmation from the remote device.

1. **REQUEST-ARGUMENT**	CAB-Service-Argument
2. **RESULT-Complex**	Status-Code
	CAB-Service-Result
1. **REQUEST-ARGUMENT**	CAB-Service-Argument
2. **RESULT-Simple**	Status-Code

The short description of CAB protocol services is listed on the following pages. To illustrate their full description for engineers and end users, two protocol services used to access network devices by operators are described fully in CAB-Confirmed-SignOn and CAB-Confirmed-SignOff services below.

ACCESS CONTROL SERVICES

CAB-Confirmed-SignOn

Service is to establish access of the operator to a desired network controller.

1. **REQUEST–ARGUMENT**	CAB-SignOn-Arg
	operator-name
	operator-password
	operator-device
	key1
2. **RESULT–Complex**	Status-Codes
	SUCCESS
	ERROR_SERVICE_TIMEOUT
	ERROR_DEVICE_BUSY
	ERROR_COMMUNICATIONS_FAILURE
	ERROR_INVALID_OPERATOR_NAME
	ERROR_INVALID_OPERATOR_PASSWORD
	ERROR_OPERATOR_SIGN_ON_LIMIT
	ERROR_OPERATOR_DEVICE
	ERROR_INVALID_ARGUMENT
	ERROR_MISC_SYSTEM_FAILURE

                                        CAB-Service-Result
                                          cab-version
                                          access-authority
                                          operator
                                          database-version

CAB-Confirmed-SignOff

The service is used to sign off from the session and to cancel the current authority level.

	1. **REQUEST–ARGUMENT**	CAB-SignOff-Arg
		operator — from sign-on
	2. **RESULT–Simple**	Status-Code
		SUCCESS
		ERROR_SERVICE_TIMEOUT
		ERROR_DEVICE_BUSY
		ERROR_COMMUNICATIONS_FAILURE
		ERROR_NO_SIGN_ON
		ERROR_INVALID_DEVICE
		ERROR_INVALID_ARGUMENT
		ERROR_MISC_SYSTEM_FAILURE

OTHER CAB PROTOCOL SERVICES

Point Status and Control

CAB-Confirmed-Set-Value: Used to control point attributes. Attributes of a single point or a cluster of points — using wildcard — can be set by this service to a desired value.

CAB-Confirmed-Inq-Value: Used to request values of single point attributes or attributes of several points, using a wildcard. For data segmentation, a hierarchical "response_to_follow_pt" for point identifiers, "response_to_follow_at" for attribute identifiers, "response_to_follow_index" for indexes such as schedule number or calendar day are included in the argument.

Virtual Terminal Services

CAB-Confirmed-Open-Terminal: Virtual terminal services are utilized to communicate to a specific controller or another OWS on a network. Depending on the network topology, individual controllers or controllers connected to the network via a CAB gateway can be addressed by an OWS residing on the same network. Each device on the network has a unique address (integer), identified in the two following services: CAB-Confirmed-Open-Term and CAB-Confirmed-Close-Term.

To access a controller from an OWS, a *CAB-Confirmed-Open-Term* protocol service is used.

CAB-Confirmed-Close-Terminal: To close the session, a CAB-Confirmed-Close-Term protocol service is used.

CAB-Unconfirmed-Send-Term is an unconfirmed protocol service used for virtual terminal sessions on the network.

Event Control

CAB-Confirmed-Sub-Event: Event subscription is related to one of the two categories:

1. Point-specific, related to real point alarms/events (i.e., high temperature alarm) of points associated with controllers connected to the network directly or via CAB-Gateways. The subscription contains the point ID and event category.

2. Nonpoint-specific, related to controllers, gateways, their communications, etc. (i.e., loss of communications between a gateway and a specific controller) on a CAB network. The subscription contains an empty point ID and event category.

Events are grouped into:

- Alarms: Critical, Cautionary, and Maintenance alarms, for point attributes changing from COS/COV to Alarm; controller-specific alarms; systems failures, detected by CAB-Gateways.
- COS/COV of digital or analog present values plus/minus their differentials.
- Miscellaneous situations determined by controllers or operators.

CAB-Confirmed-Can-Event: Used for cancellation of the above event subscription.

CAB-Confirmed-Event-Notify: Each event notification has a unique event number assigned to it by a network device. CAB network devices (controllers, OWS, etc.) subscribed to specific events will be notified when events occur. Besides the CAB network-device ID, the event number, date, and time, enumerated event type and category, event text, point ID, alarm text 1 and 2, point type, and PV are identified in the Event Notifier Argument.

CAB-Confirmed-Ack-Notify: Alarm acknowledgment notification service is used to notify subscribed OWS to critical alarm events on alarm acknowledgment by the designated operator. The notification data consists of the operator's ID acknowledging the alarm, OWS ID from which the alarm was acknowledged, the network device (controller) ID originating the alarm, and the alarm event number.

Exception Reporting Services

CAB-Confirmed-COSV-Sub: Operators can be notified on change of state (COS)/change of value (COV) by this service automatically in a more effective way than subscribing to the CAB-Confirmed-Sub-Event service.

CAB-Confirmed-COSV-Msg: Notification of COS/COV of a point attribute identified by the controller's ID, point ID, and attribute value.

CAB-Confirmed-COSV-Can: Cancellation of COS/COV notification by this service utilizing operator ID, point identifier, and an attribute ID.

Data Collection Services

CAB-Confirmed-Set-Collection: The service is used for data collection over the network. Three types of data collection functions are identified:

1. Historical data from analog points (minimum, maximum, and the average value of 5-min sample rate) are collected at a specified rate.
2. Trending data of specified attributes of analog points are collected at a specified rate.
3. Control loop plot is a collection of three attributes associated with a control loops — set point, measured value (analog input), and output position (positive feedback of an analog output).

Files of data collected in items 1 and 2 can be transferred by an operator using a CAN-Confirmed-Can-Collection; data from item 3 are transferred immediately to the requesting OWS or controller. Assignment of a unique number to every "dc_entry" enables operators to set up several data collection files of the same points and attributes at different rates (sample times) over the network.

CAB-Confirmed-Inq-Collection: "Inquire data collection entry" is to inquire specified data files, identified by dc_type, from controllers on the network.

CAB-Confirmed-Can-Collection: To cancel data collection or delete data from a data file, a "Cancel data collection entry" can be used.

CAB-Confirmed-CLP-Notify: The "control loop plot" notification service notifies the OWS on available data control loop plot.

File Services

CAB-Confirmed-Open-File: File types most commonly used in building automation systems are reflected by definition of file *types* defined in CAB protocol:

- Alarm data
- Historical data
- Trend data
- Sign-on history
- Miscellaneous (unformatted records)

Files can be opened by any of the following open modes:

- Read Only
- Read and Write
- Create (define new files)
- Update (append data to files)

In addition there are three file access modes defined:

- Sequential (starts at the beginning of the file)
- Direct (reads specified records)
- Keyed (in reference to specific key or index)

Besides "Operator" and "File-Identifier," the above must be specified in the CAB-Open-File-Argument.

CAB-Confirmed-Read-File: The service is used to access record(s) of open data files of a network device (controller or OWS).

CAB-Confirmed-Write-File: The service is used to write to a previously opened file(s).

CAB-Confirmed-Close-File: There are three defined actions for closing previously opened files:

- Keep (file is not modified)
- Delete
- Clear (clear data from a file structure)

Directory Services

CAB-Confirmed-Inq-Points: The service is used to obtain a list of point names and their associated access control levels from a specified controller on a CAB network.

CAB-Confirmed-Inq-Priv: The service is used to receive a list of CAB protocol service requests, their numbers, and associated access control authority-masks (0 to 15).

CAB-Confirmed-Inq-Systems: The service is used to obtain a list of so-called systems defined in a network control device by either unique identifiers or group of systems utilizing wildcards for their selection.

Time/Date Services

CAB-Confirmed-Set-Time: The service is used to set time in all connected network devices. Unitary or application-specific controllers connected to network controllers must set their time accordingly.

CAB-Confirmed-Req-Time: The service is used to request time and date from a network device.

Broadcast Services

CAB-Broad-Inq-Devices: The service is to inquire which network devices (OWSs, CAB-Gateways, All Devices) are active on the network.

CAB-Confirmed-Resp-Device: The service responds to previously broadcasted "inquire" request. The response includes the CAB version, type of device (OWS, CAB-Gateway, Controller, etc.), responding-address (0 to 24 octets), responding-ID-code (0 to 10 octets), device-area, area-text 1 and 2, network-unique vendor-type (number), and a printable device-text.

CAB-Broad-Mesg: The service is used to broadcast four types of messages over the network:

- Startup — the network device is coming online on the network
- Shutdown — the network device is going offline from the network
- Untransmitted alarms — request for Critical Alarms while the subscribed network device was offline
- Miscellaneous — to send messages to network devices

Save/Restore Memory Service

CAB-Confirmed-Save-Controller: The service is used by the operator to save the content of a memory of a vendor specific controller. Such backup is then available on a network utilizing File Transfer services.

CAB-Confirmed-Restore-Controller: The service is used to restore memory of a controller.

CAB-Confirm-Inq-Controller-Save: The service is to request information on the most recent save for any network controller. The service consists of all pertinent information, including four possible statuses:

- Complete — successful
- Complete — failed
- No save performed
- Save in progress

CAB-Confirmed-Inq-Controller-Restore: The service is used to request information on the most recent restore on a network controller. The service consists of all pertinent information including four possible statuses:

- Complete — successful
- Complete — failed
- No restore performed
- Restore in progress

CAB-Confirmed-Inq-Gateway-Save-Info: The service is used to request gateway save information on either full save (all files) or partial save (last changes) of files residing in CAB-Gateways.

CAB-Confirmed-Gateway-Restore-Info: The service is used to restore information in a CAB-Gateway device.

CAB-Confirmed-Gateway-Save-Rest-Done: To initiate internal data verification or file update in a gateway, an OWS can notify the gateway on a save/restore completion.

Operator Access Control Utilities Services

CAB-Confirmed-Inq-Pass: To obtain operator access information for an operator (op-name) or for all operators from a network controller or OWS, the INQ-Pass service can be used.

CAB-Confirmed-Set-Pass: The service is used to define or modify operator access information individually for each operator over the network.

CAB-Confirmed-Del-Pass: To delete operator access information individually for each operator over the network, the Del-Pass service can be used.

Miscellaneous Services

CAB-Confirmed-Are-You-There: The service polls the network controller or OWS to verify its active presence on the network. The *CAB-Are-You-There-Response* from a network device contains the *date-value*, *time-value*, and *zone*.

CAB-Confirmed-Who-Is-Signed-On: The service is used to inquire who is signed on to a remote network controller or OWS.

The *CAB-SignOns-Responds* with the *operator-name*, *signon-device*, *signon-time*, *signon-date*, *signon-address*, and *this-oper-num* (sign on number).

CAB-Confirmed-Send-Mesg: Operators on a network can communicate with each other by using the Send-Msg service.

CAB-Confirmed-Test-Device: The service is used by the operators to reboot (cold and warm boot) or initialize self-testing (RAM diagnostics, network interface, controller interface, mass storage device, etc.) of controllers or OWS residing on the network.

CAB Protocol also defines lists of ASN.1 definitions for CAB services and lists of error messages.

CAB NAMING AND ADDRESSING

Every CAB-Controller, CAB-Gateway, and OWS on a CAB network is uniquely identified by a **network address**. Network addresses (IP or Ethernet) are numerical. **Network object names** are created for the users to identify network operating work stations (OWS) or controllers on the network. The name has three sections, each with 10 alphanumeric characters:

ORGANIZATION · SUB-ORGANIZATION · NETWORK-NODE-DEVICE

Network object names in CAB standard reflect the organizational hierarchy of buildings under PWC jurisdiction.

Organization can be a company name, a region, or area of the country.

Suborganization can be a district, a group of buildings, a geographical area, school, or department.

Network-node-device would be located within the building for a certain task (i.e., building automation) and/or for specific vendor systems.

Points associated with certain network-node-devices are identified by three-level identifiers, each with 10 alphanumeric characters.

AREA · SYSTEM · POINT

Area refers to the building or area where the points are located; an area may have multiple network devices with multiple vendors, CAB-compatible controllers, OWS, or CAB-Gateways. A network device can be located only in one area.

System refers to building systems or building functions; systems are not restricted to one vendor's hardware. A system may be associated with more than one network device

(CAB-compatible controller or OWS); there could be multiple systems on any one network device.

Point refers to an actual field or pseudo-point or to a collection of data values and their attributes associated with the CAB compatible controller or CAB-Gateway. Points are unique within the entire CAB LAN; each point is associated with a particular controller or OWS on the network.

Additional identifiers for data collection and file identifiers can be added to handle data transactions in addition to the above identifiers. Two symbols — "?" and "*" — are used as wildcard selectors to identify a character or a number of characters.

CAB DATA SEGMENTATION

Data segmentation is required for services where the amount of data would accede the capacity of PDU. In such cases, the service request will include "response_to_follow" and the response will include "more_to_follow."

CAB ACCESS CONTROL

Network security and access is an important task for real-time automation systems. Operators accessing the system locally have to log-on to their PCs or controllers as per the vendor or system specific requirements. In addition, each operator accessing any network device (OWS or controller) on the CAB network has to have an access code containing the operator's name, password, and access level. The information is stored in the network device (OWS or controller) with a unique operator number. This number is used in all network services during the session (until the operator signs off, or the computer times out).

The CAB Access Control has 16 elements (0 to 15) bit mask. To access data over the CAB network, the service request, the operator access code and the point name, and attributes have to match.

Service Mask and Operator Mask AND Point Attribute Mask </> 0

CONCLUSION

The CAB protocol represents an effort of a government agency for establishing a common protocol for building automation systems installed in the Canadian government buildings. It is also an effort to standardize on automation systems which have complied with the CAB protocol specifications and are being selected in a bidding process for individually designed buildings.

Building automation system developers do not have to modify the concept (hardware and software architecture) of their systems to be compatible with the CAB protocol. However, they have to develop either CAB interfaces in their building (network) controllers, or develop CAB-Gateways for interfacing to the PWC LAN using CAB protocol.

The CAB topology assures compartmentalization of individual vendor devices in buildings or areas, making it easier for system or network troubleshooting, and division of responsibilities for individual systems performance by their respective vendors.

10 LonTalk® Protocol

Viktor Boed

CONTENTS

Echelon CORPORATION

LonTalk® protocol represents an effort of the private industry to standardize communication protocols for automation systems. Echelon Corporation started the ball rolling by bringing a Neuron® chip on the market, which was designed as a low-cost OEM VLSI chip. The Neuron chip is a trademark of Echelon Corporation. Today, over 3000 OEM vendors utilize the technology, and with four million nodes, LonTalk became one of the most recognized names in the industry.

Echelon was founded in 1988. They supply the LONWORKS® technology for building, industrial, and home control, and for automation networking. The LONWORKS MAC layer was also included into the BACnet protocol, and the LONWORKS technology became a standard (EIA 709) for integrated home systems.

Echelon offers a variety of products that enable control vendors and integrators to develop and support products that can coexist on a common network. In addition, Echelon offers network services, which enable integrators to develop and test interoperable products.

Echelon took a global approach to integration from development of a chip through network integration products to standardization and training of a cadre of authorized network integrators who could face up to the challenge of multivendor systems integration.

Neuron® CHIPS

Neuron chips are low-cost, sophisticated, very large-scale integration (VLSI) chips developed for network applications with a low-target price per chip. Neuron chips developed for large applications have an external memory interface; Neuron chips for smaller applications have 10k bytes ROM for communication protocol, operating system, and Input/Output functions accessible by application programs (Figure 10.1). Neuron chips have microprocessors dedicated to media access control (MAC) functions and protocol layer processing, and application layer and user program processing.

The chip itself is a sophisticated integrated device with built-in 24-device controllers, distributed real-time operating system, run-time libraries, three types of memories, diagnostics, watchdog timer, and a 48-bit serial number accessible by the firmware which guarantees uniqueness of every chip and every node.

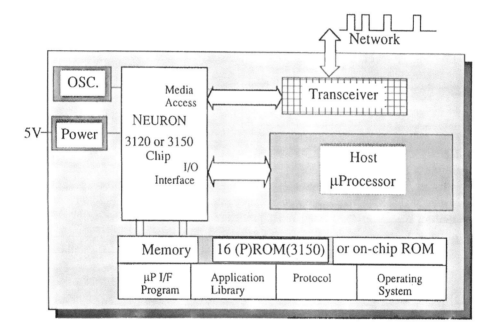

FIGURE 10.1 Example of a Motorola MC14312 Neuron® chip for small application programs. (From Echelon, Palo Alto, CA. With permission.)

LonWorks®

LonWorks networks is a family of some 80 products — hardware, software, and technology solutions aimed at systems interoperability in a multivendor control environment. These products offer solutions and flexibility that allow systems developers to develop products which can communicate on the same network, and system integrators to integrate products from various manufacturers. LonWork devices or nodes on the network communicate with each other via a LonTalk protocol.

LonTalk®

LonTalk is a protocol implemented in the firmware of a Neuron chip. It offers a collection of services used by designers to design interoperable devices.

LonMark® INTEROPERABILITY ASSOCIATION

LonMark® is an independent association founded in 1994. The association members are vendors, integrators, and end users. They are responsible for development of design standards for products based on the LonWorks technology, certification of controls, and communication products that meet the LonMark standard, and for promotion of the use of these products for control systems. The LonMark HVAC group also provides development of so-called "functional profiles" for HVAC-related hardware, such as sensors and actuators, and for different HVAC application specific controllers. The association's members are divided into three categories:

1. Sponsors are key companies, such as manufacturers of the Neuron chip, leading companies of the BAS, home automation and industrial control companies, etc. They provide not only leadership of the LonMark programs, but also financial support.
2. Partners are manufacturers participating in technical task groups, advertising, and marketing of the LonMark program.
3. Associates, such as system integrators and end users participating in LonMark activities, including field testing of new LonMark compatible products.

The membership in the association is not free; it carries a price tag assigned for each of the above categories.

LAYERS OF THE LonTalk PROTOCOL

The LonTalk protocol adheres to the seven layers of the OSI model (see Chapter 2).

LAYER 1

The LonTalk **physical layer** supports communications on various media, such as on twisted pairs, power lines, or radio frequency. LonWorks nodes are connected to the physical media (i.e., twisted pair of wires) by transceivers, which transport the packets of information. The rate at which each channel communicates depends on the design of the transceiver. Standardization of transceiver design is important to assure interoperability among devices of different manufacturers on the same network. Each transceiver must meet the design requirements and pass LonMark conformance review.

There are three Echelon transceivers currently on the market for twisted-pair communications:

1. For data rates of 78 kbps, for a bus topology with up to 64 nodes connected to a channel using 22- or 24-AWG twisted pair wiring on a typical length of 660 ft (2000 m); worst-case length of 440 ft (1330 m).
2. For data rates of 1.25 Mbps, for a bus topology with up to 64 nodes connected to a channel using 22- or 24-AWG twisted pair wiring on a typical length of 160 ft (500 m); worst-case length of 40 ft (125 m).
3. For a data rate of 39 kbps, for a bus topology and RS-485 specification, with up to 32 nodes connected to a channel using 22- or 24-AWG twisted-pair wiring on a maximum channel length of 400 ft (1200 m).

Another set of transceivers is used for power line transmission, for data rates of 2 kbps to 10 kbps with different line couplings (line to earth, line to neutral). The frequency range of transceivers is 100 kHz to 450 kHz, depending on the transceivers. The currently approved transceiver for radio frequency is for a transmission speed of 4.883 kbps and frequency ranges conforming to European standard (ETS 300220), UK standard (MPT 1329), Australian standard (RCL 1993/1), and USA FCC Part 90.

The most frequently used communications media in building automation systems are twisted pairs of wires. Integrators and end users should follow LonMark publications for transceivers conforming to LonTalk technology to keep abreast of the new technology on the market. Since transceivers are the keystone of interoperability of LONWORKS nodes, success of networking different manufacturers' products will depend on the quality and availability of transceivers.

LAYER 2

The LonTalk **link layer** contains a media access control (MAC) sublayer. The MAC sublayer is responsible for access to different physical media. In effort to provide collision avoidance, it uses a predictive p-persistent carrier-sense, multiple access collision avoidance technique based on predicting the channel load. A node accessing the channel delays its transmission to avoid collision. The LonTalk link layer provides data encoding and a 16-bit CRC error checking. The interface to the physical layer is either in a direct mode using Manchester encoding, or in a special purpose mode, in which the data are transferred serially without encoding.

There are three parameters in direct mode that have to be configured in all transceivers to communicate on the same network:

1. A preamble has to be transmitted at the beginning of each packet to synchronize the node's receiver clocks on the channel.
2. An idle period called beta 1 is to be transmitted after each packet to synchronize the end of a packet.
3. A randomizing time slot, beta 2, has to be transmitted.

LAYER 3

The LonTalk **network layer** provides connectionless network services for delivering the pockets over the variety of network configurations. The packet is delivered to one (called unicast), multiple (multicast), or all nodes (broadcast) on the network without acknowledgment. There is no retransmission or reassembly performed anywhere in the network layer.

LAYER 4

The LonTalk **transport layer** provides sequencing of incoming and outgoing messages, duplicate detection, reliable delivery, and unacknowledged repeated messages to one or more nodes on the network.

LAYER 5

The LonTalk **session layer** provides a "Request–Response" service which allows a client to communicate to a remote server.

LAYERS 6 AND 7

The LonTalk **application layer** is data oriented. The data in LonTalk protocol are of standard network variable types (SNVTs) and standard configuration parameter types (SCPTs). They are the basis for application layer interoperability among LonMark products.

LonMark objects are based on SNVTs, which includes their IDs, units, range, increment, etc.

LonTalk application programs are implemented either in the Neuron chips or in a host, such as a microcomputer, PC, etc., that contains the Neuron chip. Since the Neuron chip is present in all nodes (interoperable products) connected to the network, the network compatibility issues are straightforward.

LonMark objects (application specific or generic) are among the chief elements of interoperability among nodes connected to the same network. They must be used in design of application layer interfaces. The following generic LonMark objects are currently defined in the LonMark documentation:

- *Object type 0 — Node Object.* This allows monitoring of objects within a particular node. The object has *mandatory* (object request and object status) and optional *network variables* (time stamp, alarm, file request, file status, file position, file address). It also contains *optional configuration properties* (network configuration and maximum send time).
- *Object type 1 — Open Loop Sensor Object.* This object type can be used with any kind of sensor and digital point. The sensor object supplies data to other object types, such as to Actuator or Controller Objects.
  The mandatory network variable is the relative value from the sensor converted to appropriate engineering units and scaled.
  Optional network variables are raw data obtained from the hardware device without scaling or linearization; they are a preset function of a destination object and the feedback of a preset object.
  Optional configuration properties include the transformation of preset data from analog sensors, A/D converters, and from discrete sensors; the object's location; maximum and minimum send time for the current value; maximum and minimum range of the value; a delta value of the change of the sensor value before it is being transmitted; the offset value, which is added to the value after data translation or modification by a gain factor; invert the active polarity when a binary value is used; default output on a power off and reset; override behavior (last setting, default value) and value of a sensor when overridden; gain used in sensor calibration, translation tables x and y used for linearization and scaling; high and low limit values 1 and 2; alarm set times, alarm clear times; high and low hysteresis; alarm output inhibit time to determine inhibited alarms after the node is put online, enabled, or reset.
- *Object type 2 — Closed Loop Sensor Object.* Used in applications where more than one sensor is connected in the sensor or actuator loops. The object variables are the same as for open loop, with addition of value feedback input used to synchronize multiple sensor objects. Optional configuration properties are the same as for open loop objects.
- *Object type 3 — Open Loop Actuator Object.* This object is for applications without actuator feedback.
  The mandatory network variable is the value input.
  Optional network variables are preset input to program or preset to the output device; internal preset feedback output to synchronize objects in multiple object applications; actual position feedback output, which shows the position of the actuator.
  Optional configuration properties are location label for description of the physical location of the object; translation tables for scaling and linearization of physical movement of the actuator; input value and position feedback delays; drive time of the actuator to get from 0 to 100%; turn-off delay; default output on power failure or reset; override behavior or value; maximum receive time after the last update; high and low limit values; alarm set and alarm clear time; high and low hysteresis.
- *Object type 4 — Closed Loop Actuator Object.* Has the same values as the open loop actuator object type with an addition of a value feedback output, which synchronizes multiple objects on the network.
- *Object type 5 — Controller Object.* This object takes inputs from sensor objects and, based on the type of controller (i.e., PID controller), provides output to the actuator object. The network variables (value input and outputs and value feedback inputs and outputs) are in the receiving and output sections of the object.

SELF DOCUMENTATION

Documentation services are part of the interoperability requirements of the LonTalk protocol. They are important for documenting basic information related to the nodes of the network, which are accessible over the network or through external interface files (.FIX).

HOST-BASED NODES

Hosts use the Neuron chips as communications processors while they run the application programs on their own resources. Such nodes can be DDC vendor's controllers or PCs connected to the network. These nodes must be compatible with Neuron chip nodes.

Host-based nodes are built either with a network variable selection on the host (the host maintains the network configuration table), or the Neuron communications chip informs the host on authentication of the incoming message. In both cases, the host implements the self-documenting feature of the LonTalk protocol.

NETWORKS AND NODES

There are many ways to set up a network. In most instances, especially in facilities with existing networks, setting up a network requires adaptation of the network to existing conditions. The same implies for configuration of the network and for network topology.

Some networks have to be set up with subnetworks for individual systems, such as, for example, for HVAC control, access control, etc. Some facilities have dedicated BAS networks set up in each building, and another network connecting these individual buildings on the campus or plant level.

Regardless of the type of network, individual nodes have to be set up to communicate on the network and to share information with each other. Connecting nodes to the physical media (wiring) is insufficient — it provides the only path to communication. Logical connections have to be established for the nodes to be interoperable.

To send and receive information, each node has to have an address. The address must be unique within the network, and must contain its domain, subnet, and its node address designations. A LonTalk node can belong to up to 15 groups. A network management tool is available which provides logical connections by allocating group addresses, tracking which nodes belong to which group, assigning and reassigning network variable selectors, etc. Network variable selectors identify network variables, which allow for each node to identify its unique variables. It is up to individual DDC system design and up to the integrators to set up a network based on the facility needs.

LonTalk allows setting up subsystem gateways, with interfaces to its subsystem and also to the network (Figure 10.2). Like almost every gateway, the LonTalk gateways allow passing information, or denying access to the nodes on the subsystem. The gateway does not allow passage of information to the network nodes without such information being mapped into LonMark objects.

LonTalk nodes are composed of Neuron chips, LONWORKS transceivers, and the I/O circuitry interfacing to the field points. The family of products offered by Echelon and other LonMark-certified companies includes intelligent sensors/actuators connected directly to the LonTalk subnetwork. They also offer network devices, such as routers (i.e., Ethernet/BACnet-to-LonMark controllers) and protocol interface devices (i.e., Ethernet/MSTP router).

Since the LonTalk protocol resides in each and every Neuron chip which reside in every LONWORKS node on the network, network management functions of LONWORKS **topology** are incorporated in the protocol. Network management issues incorporated into the chip include items such as address assignment, router and bridge definition, communication service modification, traffic data collection, diagnostics, etc.

Most major DDC vendors offer their product line with a LonTalk interface. Some vendors base their communications options entirely on the LonMark protocol.

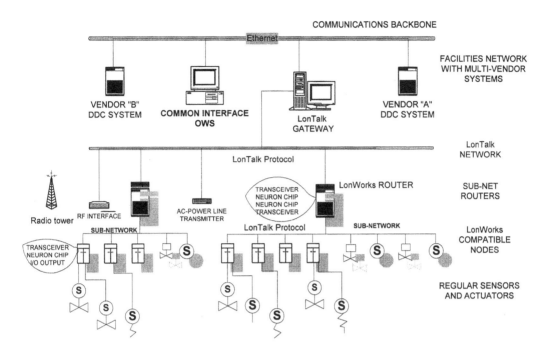

FIGURE 10.2 Block diagram of LONWORKS® connections.

BACnet COMPLIANCE

The LonTalk protocol is described in The LonTalk Protocol Specification of Echelon Corporation. ASHRAE Standard SPC 135-1995 references the LonTalk protocol. LonTalk-compatible controllers and devices conform to ISO 8802-2 Logical Link Control (LLC) Class I (Unacknowledged Connectionless-Mode) requirements. BACnet-compatible controllers may pass link service data units (LSDUs) to the LonTalk devices. BACnet DL-UNIDATA primitives have source and destination addresses and priority parameters. Source and destination addresses consist of LonTalk address, Link Service Access Point (LSAP) and message Code (MC). BACnet DL_UNIDATA is mapped into the LonTalk Application Layer Interface as:

- BACnet DL_UNIDATA.request is mapped into LonTalk msg_send request primitive
- BACnet DL_UNIDATA.indication is mapped as LonTalk msg_request primitive

SOURCES OF INFORMATION

Facilities engineers and integrators should consult the Echelon and LonMarks publications prior to deciding on implementation of LONWORKS products. Interoperability with BACnet is described in the ASHRAE SPC 135-1995 Standard.

Another source of information is the engineering bulletins of DDC vendors providing LonMark communication options.

LonTalk PROTOCOL AND ITS BENEFITS TO THE USERS

The LonTalk protocol is unique in its approach to implement a communication protocol on a single chip. By having a protocol on a silicone, compatibility issues between vendor systems utilizing Neuron chips are resolved to a great extent. The (low) cost of the Neuron chips, along with the cost of the LonMark protocol makes it attractive to building automation systems vendors. Facilities managers and engineers could benefit from the inexpensive approach to interoperability and from the simplicity of conformance testing of different vendors' systems with LonMark protocols.

The broad base of utilization of the LonTalk protocol is comforting for facilities engineers and managers. This is due to the availability of products on the market but, most important, due to availability of engineering resources.

Ongoing support of every system, whether it is a DDC or communications system, is the most important aspect for facilities managers. Systems have to be supported long after their initial implementation. Further, the cost of their operation, management, and maintenance far exceeds the initial cost for implementation. Upgrade and replacement cost is one of the major cost items during the life span of a system. Therefore, it is important not only from an engineering standpoint, but also from the financial point of view to implement systems which require minimum maintenance and minimum cost for modernization and expansions.

11 The Development of BACnet*

Ira Goldschmidt

CONTENTS

INTRODUCTION

BACnet, the standard communications protocol[1] for the HVAC controls industry, is clearly becoming the accepted alternative to the proprietary communications solutions that to date have dominated most HVAC controls installations. Its promise of interoperability has been widely anticipated for over 10 years.

As a co-author of the standard, I am often confronted with impatience regarding the pace of the standard's development and market penetration. A simple response to this concern is that interoperability in direct digital control (DDC) is a complex issue that should only be met by a carefully designed and released solution. A more cynical view is that the building design, construction, and management industry is not normally willing to participate in the learning curve of a new technology.

To make new technology palatable to the building industry, computerized controls have been sold with overblown claims and expectations. Readers with experience in first-generation, computerized energy management and DDC systems should understand this challenge and appreciate a careful transition to the industry dominance of BACnet.

BACnet products are widely available and can be found in thousands of installations. Recent articles[2,3] have documented the growing popularity of BACnet and the completion of a multivendor project at the Phillip Burton Federal Building in San Francisco (known as "450 Golden Gate"). Nevertheless, further efforts, developments, and patience are required before BACnet becomes the de facto technology in most building controls projects. This article provides insight into the challenges and complexities that were confronted in the development of BACnet. It also describes the steps remaining for full transition of the industry to BACnet. Ultimately, this story will help the reader understand that the success of the standard can only be assured through patient participation by everyone in the building industry — a corollary to "you are either part of the solution or part of the problem."

* Reprinted with permission from *Strategic Planning for Energy and the Environment*, a publication of the Association of Energy Engineers, Atlanta, GA, 1998.

THE BEGINNINGS

The growing pains in the development of computerized DDC systems in the early 1980s quickly gave way to a concern for the proprietary communications methods incorporated into these systems. DDC products from a given manufacturer could not operate within a single system with other manufacturers' products (referred to as "interoperability" in this chapter). The typical complaint leveled by users was that competitively priced additions to DDC systems could not be procured, and that these additions were limited to only those products offered by the original system's manufacturer. While these frustrations were understandable, it is important to recognize that proprietary communications were a natural result of the lack of off-the-shelf communication solutions and immaturity in digital communications technology.

Large facilities quickly became concerned about the limitations inherent in DDC systems' proprietary communications. One such user, Michael Newman at Cornell University, quickly decided to take matters into his own hands through the challenge of developing a universal "host" to Cornell's campus of multiple-manufacturer DDC systems. Additionally, some of the energy management system manufacturers that began in the 1970s and 1980s understood that dominance by industry controls giants could not be challenged without open communications. In fact, American Auto-Matrix opened a communications protocol to the industry via publication of "Public Host Protocol" in 1985. Meanwhile, many consulting engineers felt powerless to help building owners with abandoned or under-utilized DDC systems that could not be improved.

These forces led to the seminal 1987 roundtable on "Standardizing EMS Protocols"[4] organized by *Energy User News* in New York City. This roundtable highlighted the coincidental announcement[5] that Mike Newman would chair an ASHRAE committee to develop a standard protocol. These events drew support for the ASHRAE committee from those that attended or found the *Energy User News* articles compelling.

A few consulting engineers, like this author, were drawn to the committee with a "revenge of the nerds" goal, and were hoping to use the standard on projects that were just entering design. The committee came together, optimistic that, with cooperation from all involved, the standard could be completed in a year. Unfortunately, Mike Newman's prediction that if "...cooperation is less than complete, it could take forever" was closer to the truth.

GATHERING MOMENTUM

Early meetings of the committee quickly led to the realization that developing a standard communications protocol was a technological and political challenge well beyond our initial optimism. We quickly discovered that concurrent and interdependent work would be required on a number of issues, including:

- **Terminology** — Agreement on the definition of common terms such as "host," "download," and "warm-start" was needed to avoid the "tower of Babel" besetting committee meetings.
- **Scope** — Should the standard apply to host-to-controller communications, controller-to-controller communications, or both? Should it apply to all types of controllers, including terminal/zone controllers?
- **Services** — Should the protocol support system startup and configuration tasks (e.g., programming) in addition to operations tasks (e.g., viewing point values)? Is changing a control set point a configuration or an operation task?
- **Data** — Should we define complex data structures based on HVAC equipment (e.g., chillers and boilers) or more simple structures based on generic engineering data (e.g., temperatures)?

- **Choice** — Should the protocol allow multiple ways to communicate the same data?
- **Extension** — Should the standard be allowed to be extended with proprietary innovations?
- **Physical path** — Should existing LAN technologies (e.g., Ethernet) be adopted and/or should new LAN technologies be developed (e.g., based on EIA-485)?
- **Encoding** — How should messages be efficiently encoded into the "0s" and "1s" required of a digital communications system?
- **Structure** — Should the standard be modeled after the new ISO "Open Systems Interconnection (OSI)" model?

In addition to the above technical issues, the politics inherent in gaining consensus from competing manufacturers, some of whose representatives appeared to be threatened by the goals of the committee, led some of us to wonder if we had cashed a check on an account that could never be opened.

Fortunately, after a few initial meetings, the committee started to make some important choices, including:

- Use an object-oriented approach to define a small set of data structures common to DDC systems, e.g., points, schedules, alarms
- Provide choices in services and physical paths (i.e., LANs) that both allow, simple controllers to operate on the network at a reasonable cost and bigger controllers to operate efficiently
- Divide the committee into three major task groups to address the distinct components of the standard: application services, object types and properties, and encoding
- Define terminology only when absolutely needed
- The standard should not support services for product configuration; the concern was that this would stifle the creativity and competition in the design of DDC products
- Follow a subset of the OSI model to avoid unnecessary cost and complexity

Meanwhile, other efforts to create open/standard protocols in the late 1980s, notably by the Intelligent Buildings Institute and Public Works Canada, put pressure on the need for ASHRAE to move ahead.

THE CHALLENGE

As the committee's efforts continued into the 1990s, it became obvious that each meeting would devote significant time on revisiting old issues. It was not always clear whether this constant rehashing was due to opposition to the standard or just a lack of understanding. Ironically, while it was often tempting to give up in frustration, this constant prodding and reevaluation proved to test the soundness of our decisions and would lead to a better standard.

We continued to refine the choices made earlier in the standard's development. In particular, the decision to develop an EIA-485 LAN technology — later known as "MS/TP" — meant that extensive protoyping would be required. In the end, this effort required several years and extensive offline efforts by members skilled at electronic design.

To keep consensus, optional data parameters were defined, and choices in implementing services were allowed. It was understood that these options and choices would make interoperability difficult, but we expected that the market would constrain the use of the standard to achieve interoperability. I'm not sure if the committee fully understood the ramifications of this decision.

Early in the development of the standard, we became aware of a new product offering called LonTalk®. Its message delivery functions (not including its applications services and data structures) appeared to be an off-the-shelf alternative to MS/TP (i.e., a low cost/low speed LAN). However, concerns over its proprietary origins (it was developed and largely controlled by Echelon, Inc., Palo Alto, CA) meant that it would not be included in the first public review of the standard in 1991.

Eventually, pressure from manufacturers making investments in the development of LonTalk-based products led to a showdown on the issue. Committee members not committed to the use of LonTalk were concerned that its growing popularity was more a result of big marketing dollars than the benefit it could provide to BACnet. The issue of including LonTalk as a LAN technology within BACnet was passed prior to the third public review of the standard in 1995. There appeared to be a deadlock on the potential appeal of BACnet by Echelon, at the final vote for adoption, which led to an observer's remark that some committee members held their noses while voting "yes."

Unfortunately, the large disparity between the services and data structures of BACnet and LonTalk means that a BACnet system will never interoperate with a full LonTalk system without the use of a gateway. (For an excellent discussion on gateway issues, see Reference 6.)

Less controversial, the committee also included IEEE 802.3 (the standardized version of Ethernet) and ARCNET as high-speed LAN choices, and developed a direct/modem connection technology called PTP (for "point-to-point").

The question of whether a standard protocol could constrain DDC product design continually brought heated discussions. Eventually, it became clear that some constraint was inevitable, given the goals of BACnet. Committee member representatives from some manufacturers were not overjoyed with this prospect, because of the cost to redesign their products to implement the standard. To help soften the blow, we tried to develop models for such functions as alarming and scheduling that drew on some of these manufacturers' current philosophies. Again, to gain consensus, options and choices were also included.

A realization that occurred just shortly before the completion of the standard's first draft was that modem DDC systems required routing functions to allow for a connection of multiple DDC networks. This realization led to expanding the standard's use of the OSI model to include a network layer made up of simple functions created by the committee. This decision would help pave the way for BACnet's future use on the Internet.

One of the final efforts before the release of the standard was to develop a method to facilitate the specification of BACnet-based systems. It was understood that an intimate knowledge of the 500-page standard could not be expected of the consulting engineering community. Therefore, the "Conformance and Specification" clause was written as an attempt to provide everything a consulting engineer would need to know to specify BACnet. It was never the intent of this section to constrain BACnet's myriad choices and options to provide interoperability.

Of course, the wisdom in the above decisions is not yet fully proven. However, experience with other successful standards and de facto standards (e.g., the PC) has shown that an excessive focus on perfection is not necessary, and may even be the kiss of death.

BIRTH OF A STANDARD

It would have been easy to forever find reasons to delay the release of the standard, especially with a committee composed of engineers, some of whom were apparently opposed to BACnet. However, it became clear by 1991 that the standard must be constrained to the core issue of a communications protocol to hasten its completion. In particular, we chose to leave a number of peripheral issues for future efforts. These issues included development of a method for testing conformance to the standard, and the selection of an organization for managing the certification of BACnet products. It was understood that after release of the standard, these issues would need attention before BACnet could become fully viable.

Two other key events served to hasten BACnet's release:

1. Product development and testing — A standard that defines a complex set of rules governing digital communications cannot be developed exclusively on paper. At some point, the ideas must be prototyped in real devices as a true test of soundness. The "BACnet Interoperability and Testing Consortium" was organized by the National Institute of Standards and Technology to provide an environment where manufacturers could test product prototypes. Unfortunately, participation in this consortium was halfhearted until a commitment to complete the standard was made.
2. The Trane Company chose to market a BACnet-compatible product well before completion of the final version of the standard. This gambit probably helped to generate market interest and to convince other manufacturers that any further delays might relinquish a major advantage to Trane.

As an ANSI standards body, ASHRAE is bound by the rules of public review and comment. From the time of BACnet's first published draft it took 4 years, 3 public reviews, and the individual resolution of 741 comments to gain approval of formal publication of the standard in 1995. This process helped make BACnet stronger than any proprietary protocol could ever hope for, but delayed its release to the point where it could have died on the vine.

UNFINISHED BUSINESS

Completion of the BACnet standard merely marked the start of a "re-tooling" of the controls contracting industry for delivery of truly integrated building automation solutions." For this "retooling" to occur, the committee must complete some unfinished business that has been in the works for a number of years, including:

- Development of a method to test conformance to the standard
- Selection of a certification agency
- Redesign of the "Conformance and Specification" clause to better ensure interoperability

This last effort is the cause of much current controversy with the standard. As discussed earlier, the "Conformance and Specification" clause was never expected to ensure interoperability. It was always hoped that the market would constrain the choices and options within BACnet to the degree needed to provide interoperability. However, manufacturers are gun-shy of releasing products that may not interoperate with other manufacturer's BACnet products. The committee never foresaw this "Catch-22."

This quandary is due to the insistence by most manufacturers during the development of the standard to include choices and options — the very choices and options that are the cause of this interoperability challenge. So, the committee is now completing efforts to rewrite the "Conformance and Specification" clause to provide the degree of constraint needed to assure interoperability. Ironically, these constraints will undoubtedly make obsolete many of the sacred cows that were originally included in BACnet for the purpose of achieving consensus.

With the completion of the above efforts expected before the end of the year, the retooling of the controls industry will involve a number of possible aspects, including:

- **Completion** of carefully-monitored projects that involve the conformance and certification methods, and the new "Conformance and Specification" clause
- **Education** of the industry, especially to help avoid the pitfalls of unrealistic expectations and to clarify confusion about the relationship of BACnet and LonTalk

- **Production** by new manufacturers of niche hardware (e.g., routers and gateways) and software (e.g., specialty operator interfaces)
- **Transition** of controls contractors from single-manufacturer providers to "systems integrators"

PARTING WORDS

BACnet will continue its path to market dominance because it represents a comprehensive consensus of industry ideas. This dominance can be hastened through support of manufacturers that are upgrading their products to comply with the standard. This support could result in your participation in single-manufacturer installations that use these products, while carefully avoiding the pitfalls of premature or overblown expectations of interoperability. The experience, profits, and good publicity that come from these scaled-down BACnet projects will contribute to the day when full scale, interoperating BACnet installations are the norm.

REFERENCES

1. ANSI/ASHRAE Standard 135-1995: BACnet — A Data Communication Protocol for Building Automation and Control Networks, ASHRAE, 1995.
2. Applebaum, M.A. and Bushby, S.T., BACnet's first large-scale test, *ASHRAE J.,* July 1998, 23.
3. Gahran, A., BACnet on Duty: New BAS frontiers for end users, *Energy User News,* March 1998, 16.
4. Mullin, R., Standardizing EMS protocols: an EUN panel discussion, *Energy User News,* February 23, 1987, 4.
5. Racanelli, V., ASHRAE forms group to seek standard EMS protocol, *Energy User News,* January 26, 1987, 1.
6. Bushby, S. T. Communication gateways: friend or foe?, *ASHRAE J.,* April, 1998, 50.

12 A Data Communication Protocol for Building Automation and Control Networks: BACnet

Viktor Boed

CONTENTS

INTRODUCTION

This chapter provides essential information for facilities engineers and managers who are not communications experts or system integrators. The review of the BACnet protocol should enable building managers and engineers to understand the basic features of the protocol, which could help them in the review, analysis, and selection of building automation systems (BAS) communication options for their installations. The information in this chapter is based on the BACnet standard and is condensed and modified for the above-stated purposes.

It is not the intention of this chapter to provide an in-depth analysis or description of the protocol for designers of communication protocols or integrators of BAS systems. For that, they should consult the ANSI/ASHRAE Standard 135-1995 and its subsequent revision.

ANSI/ASHRAE STANDARD 135-1995

Under the auspices of the American Society of Heating, Refrigerating, and Air-Conditioning Engineers (ASHRAE) a Voluntary Consensus Standard was developed and approved by the American National Standards Institute in December 19, 1995. ASHRAE sponsored a standardization committee (SPC-135) with a task to formulate a protocol that would be acceptable by building automation vendors as well as end users. The committee was established in the mid 1980s. Voting and nonvoting members of the committee were primarily from the controls industry, with participation by members of the consulting and specifying engineering community, facilities, and the so-called general interest groups. The protocol went through several public reviews and was approved by ANSI and ASHRAE in 1995 as ASHRAE Standard 135.

BACnet OBJECTIVES

The main objective of the standard is definition of data communications services and protocols for HVAC control equipment for buildings. The standard also can be applied for integration of other computerized building systems currently used in conjunction with BAS, such as security, fire detection, lighting control, and other systems. Due to BACnet's adherence to the ISO model, interfaces to other networks can be configured utilizing protocols based on the ISO model.

BASIC CHARACTERISTICS OF THE BACnet

The standard defines a set of entities called **objects**, which can be accessed from the network by writing to and reading their **attributes**, called **properties** of the BACnet defined objects. Objects are defined in network-accessible computer equipment, such as operator work stations (OWS), terminals, and BAS controllers. All network devices are considered functional peers, and use peer-to-peer communications, with exception of network devices using the master-slave/token passing (MS/TP) communication option on the second, Data Link layer, of the BACnet architecture. BACnet also defines sets of protocol services such as for virtual terminals, remote devices, object access, alarm and event services, network security, and data encoding. Depending on the ability of the network devices to support BACnet objects and/or carry out BACnet services, network devices have to conform to one of the **six BACnet conformance classes**. Since not all BAS controllers are designed for the highest conformance level, specifying engineers and end users have to understand these conformance classifications in order to achieve the desired and required interoperability for their applications.

THE BACnet MODEL

The BACnet is a collapsed model of the OSI format, utilizing *four* out of seven layers of the OSI model (Figure 12.1).

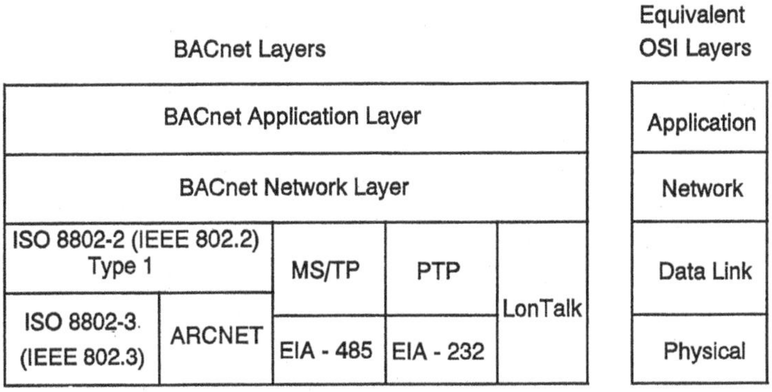

BACnet Layers					Equivalent OSI Layers
BACnet Application Layer					Application
BACnet Network Layer					Network
ISO 8802-2 (IEEE 802.2) Type 1		MS/TP	PTP	LonTalk	Data Link
ISO 8802-3 (IEEE 802.3)	ARCNET	EIA - 485	EIA - 232		Physical

FIGURE 12.1 BACnet collapsed architecture and equivalent ISO layers. (From *BACnet, a Data Communication Protocol for Building Automation and Control Networks*, ANSI/ASHRAE Standard 135-1995. With permission.)

THE BACnet LAYERS

The first (Physical) and second (Data Link) layers, the two lowest layers of the OSI model, match the same layers of the BACnet architecture and correspond to four options of the BACnet protocol:

Option 1: Logical link control (LLC), standard IEEE 802.2 type 1 (unacknowledged connectionless service), in combination with carrier sense multiple access with collision detection (CSMA/CD), medium access control (MAC), standard IEEE 802.3. CSMA/CD standard IEEE 802.3 is an access method used by the Ethernet (10 megabits per second — Mbps) protocol.

Option 2: Logical link control (LLC), standard IEEE 802.2, in combination with ARCNET (2.5 Mbps). ARCNET was originally developed as a proprietary protocol by the Data Point Corporation (Standard ATA/ANSI 878.1) for baseband LANs with token-passing access.

Option 3: Master-slave/token passing (MS/TP) protocol designed for BAS, interfacing directly to the third (network) layer, in combination with RS- 485 EIA standard.

Option 4: a hardwired or dial-up serial, asynchronous point-to-point data transmission, in combination with the RS-232 EIA data transmission standard.

Option 5: the LonTalk protocol designed by the Echelon Corporation, now supported by LonMark organization, is interfacing directly to the network layer. All lower layers are defined in the LonTalk protocol.

The above five options provide the services associated with the second layer, such as grouping of data into frames and packets, addressing, access to the medium, flow control, and some error recovery. In addition, it also provides OSI transport layer services associated with data transmission required for BAS, such as point-to-point delivery of messages, segmentation, sequence control, and other necessary services.

The third (BACnet Network) layer. The BACnet architecture allows for design of a single or multiple networks. In a case of a single network, most network layer functions are included in the data link layer. In the case of multiple networks interconnected by bridges or repeaters (with a single local address), BACnet defines the network layer header with essential addressing and control information. For multiple networks, the network layer provides translation of global to local addresses, routing of messages (one logical path only) through the interconnected networks, sequencing, flow control, error control, and multiplexing for different network types. The network layer is necessary for recognizing the differences between local and global addresses on networks with different MAC options, and routing them via proper links.

The fourth (BACnet Application) layer provides the actual interfaces with the application programs of the BAS device providing control, monitoring, and other building automation functions. It also provide functions otherwise associated with full OSI model layers, such as:

- Message segmentation and flow control, sequence control of the segmentation process (assembly and reassembly of segmented messages), provided by the transport layer of the full OSI model
- Synchronization checkpoints and their resetting in case of error, provided by the session layer of the OSI model, is included in the BACnet application and network layers, due to the short nature of most messages, such as alarms, changes of state, download of set-points, etc., with the exception of longer messages during up/download of programs into controllers
- Translation of data from the application layer to sequences of octets for the lower layers (transfer syntax) provided by the presentation layer of the OSI model is included in the application layer of the BACnet protocol

BACnet OBJECTS AND THEIR PROPERTIES

GENERAL

One of the lasting contributions of the BACnet standardization committee to the industry is the definition of commonly used objects (previously called "points") and their attributes (previously called "characteristics") for BAS. These sets of objects and attributes provide clear definitions and common understanding of their meaning, regardless of systems and manufacturers. If properly used, they provide a common language for all involved, from vendors to end users. End users can use the BACnet objects (and attributes) to evaluate or compare DDC systems for a given application and communication option. Because of the above, this chapter lists most of the BACnet objects and their properties. It also provides examples and explanations for facilities managers and engineers who are not proficient with BAS communications. The provided tables, along with conformance classes, can be used to evaluate individual DDC bids.

To exchange information over the network, the data have to be visible over the network. BACnet models different **object** types commonly used in the controls industry and defines their **properties**. Application layer services access the **standard** BACnet objects and manipulate their properties.

Objects are identified by their unique "key" property, called **object identifier**. In addition, each object type is also identified by its **property datatype** and **conformance codes**. Not all BACnet object types have to be supported by the connected BAS system.

The following are the BACnet conformance codes assigned to every property identifier:

- **O** = Optional (the property does not have to be in all BACnet objects of that particular type.) For example, an analog input object's property, "*Description*" — a character string describing, say, an "air handling unit one fan discharge temperature" — does not have to be defined in all connected controllers.
- **R** = Required and readable by BACnet services (the property is required to be present in all BACnet objects of that particular type). For example, an analog input object's property, "*Object_Name*" — a character string providing an acronym or tag name, say,

"*AHU1FDT*" for "air handling unit one fan discharge temperature" — is required to be defined in all connected controllers.

- **W** = Required, readable and writable by BACnet services (a property which is *required* to be present in all connected BACnet standard objects; its value can be changed by one or more WriteProperty services defined in the BACnet standard). For example, a "*Present_Value*" of an analog output object of a damper control is a real value, say, 50.0%, to which the damper can be commanded by any of the BAS controllers connected to the BACnet.

Nonstandard object types connected to the BACnet have to support at a minimum the following properties:

Object_Identifier, Object_Name, and Object_Type.

BACnet standard objects must support all properties (required and optional) as defined in the standard, and must return the datatype defined for that particular property. Exceptions must be detailed in the "Protocol Implementation Conformance Statement."

BACnet lists all supported standard object types under "Modeling Control Devices as a Collection of Objects" in the ASHRAE Standard 135-1995.

For purposes of this publication, and in an effort to make the concept more understandable to engineers who are not communications or protocol experts, the following BACnet object examples are reprinted and modified from the standard. In addition, there are comments and explanations for properties to make the examples more understandable to application, consulting, and facilities engineers.

Let us look at an example of an air handling unit controlled by a DDC system. The DDC system contains the application program and the operating logic. The following objects are defined as BACnet objects and are visible over the BACnet network to other systems connected to the same network.

ANALOG INPUT OBJECT

Analog input objects, such as temperature, pressure, relative humidity, and other inputs from field sensors (i.e., current, voltage, resistor inputs) are among the most commonly used objects in DDC applications. If such objects are to be made visible to other nodes (controllers, computers, etc.) on the BACnet-compatible network, their properties have to be defined. The following Figure 12.2 and Tables 12.1 and 12.2 illustrate and describe an example of a fan discharge temperature (FDT) analog input object.

ANALOG OUTPUT OBJECT

Analog output objects are most commonly used to command field analog devices, such as, for example, actuators for valve or damper control to their desired positions. Depending on the field device, the DDC analog output can be current, voltage, or other. If such objects are to be made visible to other nodes (controllers, computers, etc.) on the BACnet-compatible network, their properties have to be defined. The following Figure 12.3 and Table 12.3 illustrate and describe an example of a cooling coil valve control (CCVC) definition of an analog output object.

ANALOG VALUE OBJECT

Analog value objects are the results of calculations residing in the memory of a computer or controller connected to the BACnet-compatible network. In the past, we used to refer to them as "calculated points," "pseudo-points," "software points," etc. They do not have associated field

AIR HANDLING UNIT AH1

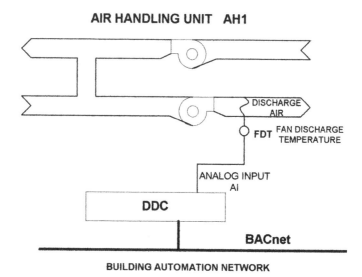

FIGURE 12.2 AI — Fan discharge temperature.

TABLE 12.1
BACnet Analog Input Object — Fan Discharge Temperature (FDT)

Property Identifier	Property Datatype	Property Datatype Example	Conformance Code	Comments and Examples
Object_Identifier	BACnetObjectIdentifier	X'00000001'	R	Key property, 1st analog input
Object_Name	CharacterString	"AH1FDT"	R	Use site-specific naming conventions
Object_Type	BACnetObjectType	Analog_Input	R	Description of object type
Present_Value	REAL	55.4	R, W	Actual reading; W-when Out_Of_Service = True
Description	CharacterString	"Fan Discharge Temperature"	O	Use site-specific naming conventions
Device_Type	CharacterString	"1000 Ohm RTD"	O	Use actual field sensor type
Status_Flags	BACnetStatusFlags	{False, False, False, False}	R	4Flags: {In_Alarm, Fault, Overriden, Out_Of_Service}; Status: True (1), False (0)
Event_State	BACnetEventState	Normal	R	To determine if the object has an associated event state
Reliability	BACnetReliability	No_Fault_Detected	O	Values: {No_Fault_Detected, No_Sensor, Over_Range, Under_Range, Open_Loop, Shorted_Loop, Unreliable_Other}

TABLE 12.1 (continued)
BACnet Analog Input Object — Fan Discharge Temperature (FDT)

Property Identifier	Property Datatype	Property Datatype Example	Conformance Code	Comments and Examples
Out_Of_Service	Boolean	False	R	True=the sensor is out of service; False=sensor in service
Update_Interval	Unsigned	20	O	In sec when not overriden or out of service
Units	BACnetEngineeringUnits	Degrees — Farenheit	R	Enumerated vale 1–141 assigned by BACnet
Min_Pres_Value	REAL	–50	O	The lowest range of the sensor
Max_Pres_Value	REAL	250	O	The maximum range of the sensor
Resolution	REAL	0.1	O	The smallest recognizable value in engineering units
COV_Increment	REAL	0.2	O, R	The min change in value that will cause COVNotification R=if the object supports COV reporting
Time_Delay	Unsigned	15	O, R	Defined in seconds; the analog value to remain outside of its limits prior to reporting a To-OFF Normal or return To-Normal condition R=if intrinsic reporting is supported
Notification_Class	Unsigned	3	O, R	BACnet defines a Notification Class object with properties for event notification over the BACnet R=if intrinsic reporting is supported.
High_Limit	REAL	63	O, R	FDT>63degF will generate High_Limit event R=if intrinsic reporting is supported. See Table 12.2 for conditions generating High_Limit
Low_limit	REAL	53	O, R	FDT<53degF will generate Low_Limit event R=if intrinsic reporting is supported. See Table 12.2 for conditions generating Low_Limit

TABLE 12.1 (continued)
BACnet Analog Input Object — Fan Discharge Temperature (FDT)

Property Identifier	Property Datatype	Property Datatype Example	Conformance Code	Comments and Examples
Deadband	REAL	2	O, R	A range the PV will remain in during normal conditions R=if intrinsic reporting is supported
Limit_Enable	BACnetLimitEnable	{True, True}	O, R	Two flags to enable and disable High and Low Limit reporting R=if intrinsic reporting is supported
Event_Enable	BACnetEventTransitionBits	{True, False, True}	O, R	Three flags: TO-OFF NORMAL, TO-FAULT, TO-NORMAL R=if intrinsic reporting is supported
Acked_Transitions	BACnetEventTransitionBits	{True, True, True}	O, R	Three flags: TO-OFF NORMAL, TO-FAULT, TO-NORMAL R=if intrinsic reporting is supported
Notify_Type	BACnetNotifyType	Event	O, R	Events or Alarms R=if intrinsic reporting is supported.

TABLE 12.2
BACnet Definition of Conditions to Generate TO–OFFNORMAL and TO–NORMAL Events

Limit	TO–OFFNORMAL	TO–NORMAL
High_Limit	PV>High_Limit for set Time_Delay, and HighLimitEnable = Limit_Enable, and TO-OFFNORMAL = Event_Enable	PV<High_Limit — Deadband for set Time_Delay, and HighLimitEnable = Limit_Enable, and TO-NORMAL = Event_Enable
Low_Limit	PV<Low_Limit for set Time_Delay, and LowLimitEnable = Limit_Enable, and TO-OFFNORMAL = Event_Enable	PV>Low_Limit — Deadband for set Time_Delay, and LowLimitEnable = Limit_Enable, and TO-NORMAL = Event_Enable

hardware (i.e., sensor). Such object may be, for example, enthalpy calculated from analog input objects of dry bulb and wet bulb sensors, average temperature calculated from several zone temperature sensors, etc. If such objects are to be made visible to other nodes (controllers, computers, etc.) on the BACnet-compatible network, their properties have to be defined. The following Figure 12.4 and Table 12.4 illustrate and describe an example of an average zone temperature (AVZT) analog value object.

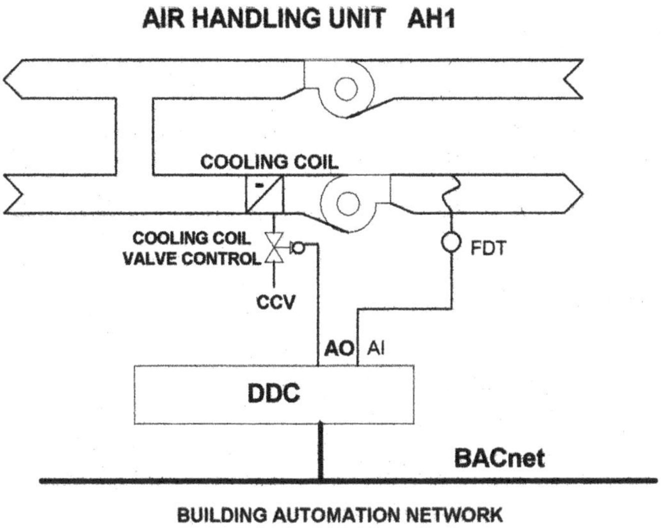

AIR HANDLING UNIT   AH1

FIGURE 12.3   AO — Cooling coil valve control.

TABLE 12.3
BACnet Analog Output Object — Cooling Coil Valve Control (CCVC)

Property Identifier	Property Datatype	Property Datatype Example	Conformance Code	Comments and Examples
Object_Identifier	BACnetObjectIdentifier	X'00000001'	R	Key property, 1st analog output
Object_Name	CharacterString	"AH1CCV"	R	Use site-specific naming conventions
Object_Type	BACnetObjectType	Analog_Output	R	Description of Object type
Present_Value	REAL	60.0	W	Actual cooling coil valve commanded position
Description	CharacterString	"Cooling Coil Valve Position"	O	Use site-specific naming conventions
Device_Type	CharacterString	"4-20 mA"	O	Use actual field actuator value, i.e., 40% OPN
Status_Flags	BACnetStatusFlags	{False, False, False, False}	R	4Flags: {In_Alarm, Fault, Overriden, Out_Of_Service};Status: True (1), False (0)
Event_State	BACnetEventState	Normal	R	To determine if the object has an associated event state
Reliability	BACnetReliability	No_Fault_Detected	O	Values: {No_Fault_Detected, Open_Loop, Shorted_Loop, Unreliable_Other}

TABLE 12.3 (continued)
BACnet Analog Output Object — Cooling Coil Valve Control (CCVC)

Property Identifier	Property Datatype	Property Datatype Example	Conformance Code	Comments and Examples
Out_Of_Service	Boolean	False	R	False = The valve is in service
Units	BACnetEngineeringUnits	Percent	R	enumerated vale 1–141 assigned by BACnet
Min_Pres_Value	REAL	0.0	O	The lowest range of the actuator
Max_Pres_Value	REAL	100	O	The maximum range of the actuator
Resolution	REAL	0.1	O	The smallest recognizable value in engineering units
Priority_Array	BACnetPriorityArray	{NULL,NULL,NULL NULL,60.0..,NULL}	R	NULL = no existing command at that priority
Relinquish_Default	REAL	100.0	R	Default value, when Priority_Array = NULL value
COV_Increment	REAL	0.2	O, R	The min change in value that will cause COV Notification; R=if the object supports COV reporting
Time_Delay	Unsigned	10	O, R	Defined in seconds; the analog value to remain outside of its limits prior to reporting a To-OFF Normal or return To-Normal condition R=if intrinsic reporting is supported
Notification_Class	Unsigned	2	O, R	BACnet defines a notification class object with properties for event notification over the BACnet R=if intrinsic reporting is supported.
High_Limit	REAL	101	O, R	CCV command>100% will generate High_Limit event R=if intrinsic reporting is supported. See Table 12.2 for conditions generating High_Limit

TABLE 12.3 (continued)
BACnet Analog Output Object — Cooling Coil Valve Control (CCVC)

Property Identifier	Property Datatype	Property Datatype Example	Conformance Code	Comments and Examples
Low_limit	REAL	−99	O, R	CCV command<0% will generate Low_Limit event R=if intrinsic reporting is supported. See Table 12.2 for conditions generating Low_Limit
Deadband	REAL	2	O, R	A range the PV will remain in during normal conditions R=if intrinsic reporting is supported
Limit_Enable	BACnetLimitEnable	{True, True}	O, R	Two flags to enable and disable High and Low Limit reporting R=if intrinsic reporting is supported
Event_Enable	BACnetEventTransitionBits	{True, False, True}	O, R	Three flags: TO-OFF NORMAL, TO-FAULT, TO-NORMAL R=if intrinsic reporting is supported
Acked_Transitions	BACnetEventTransitionBits	{True, True, True}	O, R	Three flags: TO-OFF NORMAL, TO-FAULT, TO-NORMAL R=if intrinsic reporting is supported
Notify_Type	BACnetNotifyType	Event	O, R	Events or Alarms R=if intrinsic reporting is supported

BINARY INPUT OBJECT

Binary input objects are typically inputs from contacts, such as from relays, auxiliary contacts of motor starters, end-switches, etc. They indicate the state of a controlled device, such as a pump or fan, or positions of, say, dampers, actuators, etc. (for example, an open position of a two-position damper). They could also be inputs from safety devices, such as freeze-stats, low pressure switches, etc. Two states of binary inputs are either 0 or 1, described also as Open/Closed, On/Off, Normal/Alarm, etc. The state of binary input objects in BACnet definitions is either ACTIVE (1, the equipment is On, Closed, Normal, etc.), or INACTIVE (0, the equipment is Off, Closed, Alarm, etc.). These logical states, however, can be reversed (in software definition or associated hardware). For definitions, see properties for "Present_Value" and "Polarity" in the table below.

If such objects are to be made visible to other nodes (controllers, computers, etc.) on the BACnet-compatible network, their properties have to be defined. Figure 12.5 and Table 12.5 illustrate and describe an example of a supply fan status (SFS) definition of a binary input object.

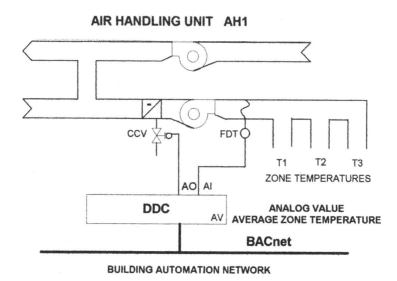

FIGURE 12.4 AV — Average zone temperature.

TABLE 12.4
BACnet Analog Value Object — Average Zone Temperature (AVZDT)

Property Identifier	Property Datatype	Property Datatype Example	Conformance Code	Comments and Examples
Object_Identifier	BACnetObjectIdentifier	X'00000001'	R	Key property, 1st analog value input
Object_Name	CharacterString	"AH1AVZT"	R	Use site-specific naming conventions
Object_Type	BACnetObjectType	Analog_Value	R	Description of object type
Present_Value	REAL	56.5	W	Actual value in engineering units
Description	CharacterString	"Average Zone Temperature"	O	Use site-specific naming conventions
Status_Flags	BACnetStatusFlags	{False, False, False, False}	R	4Flags: {In_Alarm, Fault, Overriden, Out_Of_Service}; Status: True (1), False (0)
Event_State	BACnetEventState	Normal	R	Determines if the object has an associated event state
Reliability	BACnetReliability	No_Fault_Detected	O	Values: {No_Fault_Detected, Over_Range, Under_Range, Unreliable_Other)

TABLE 12.4 (continued)
BACnet Analog Value Object — Average Zone Temperature (AVZDT)

Property Identifier	Property Datatype	Property Datatype Example	Conformance Code	Comments and Examples
Out_Of_Service	Boolean	False	R	If True=the PV can't be modified by the local software
Units	BACnetEngineeringUnits	Degrees — Farenheit	R	enumerated vale 1–141 assigned by BACnet
Priority _Array	BACnetPriorityArray	Null	R	If Priority _Array is present then Relinquish_Default must be also present
Relinquish_Default	REAL	60.0	R	If Priority _Array is present then Relinquish_Default must be also present
COV_Increment	REAL	0.2	O, R	The min change in value that will cause COV Notification; R=if the object supports COV reporting
Time_Delay	Unsigned	5	O, R	Defined in seconds; the analog value to remain outside of its limits prior to reporting a To-OFF Normal or return To-Normal condition R=if intrinsic reporting is supported
Notification_Class	Unsigned	3	O, R	BACnet defines a notification class object with properties for event notification over the BACnet R=if intrinsic reporting is supported.
High_Limit	REAL	60	O, R	AVZT>60 degF will generate High_Limit event R=if intrinsic reporting is supported. See Table 12.2 for conditions generating High_Limit

TABLE 12.4 (continued)
BACnet Analog Value Object — Average Zone Temperature (AVZDT)

Property Identifier	Property Datatype	Property Datatype Example	Conformance Code	Comments and Examples
Low_limit	REAL	50	O, R	AVZT<50 degF will generate Low_imit event R=if intrinsic reporting is supported. See Table 12.2 for conditions generating Low_Limit
Deadband	REAL	2	O, R	A range the PV will remain in during normal conditions R=if intrinsic reporting is supported
Limit_Enable	BACnetLimitEnable	{True, True}	O, R	Two flags to enable and disable High and Low Limit reporting R=if intrinsic reporting is supported
Event_Enable	BACnetEventTransitionBits	{True, False, True}	O, R	Three flags: TO-OFF NORMAL, TO-FAULT, TO-NORMAL R=if intrinsic reporting is supported
Acked_Transitions	BACnetEventTransitionBits	{True, True, True}	O, R	Three flags: TO-OFF NORMAL, TO-FAULT, TO-NORMAL R=if intrinsic reporting is supported
Notify_Type	BACnetNotifyType	Event	O, R	Events or alarms. R=if intrinsic reporting is supported

BINARY OUTPUT OBJECT

Binary output objects are typically binary outputs from the DDC to relays, motor starters, etc. They control devices, such as pumps or fans, or command dampers to open or closed positions (for example, a minimum air damper). Two states of binary outputs are either 0 or 1, described also as Open/Closed, On/Off, Start/Stop, etc. The state of binary output objects in BACnet definitions is either ACTIVE (1, the equipment is commanded to Start, turn On, to Open position, etc.), or INACTIVE (0, the equipment is commanded to Stop, turn Off, to Close position, etc.). These logical states, however, can be reversed (in software definition or associated hardware). For definitions see properties for "Present_Value" and "Polarity" in Table 12.6. If such objects are to be made visible to other nodes (controllers, computers, etc.) on the BACnet-compatable network, their properties

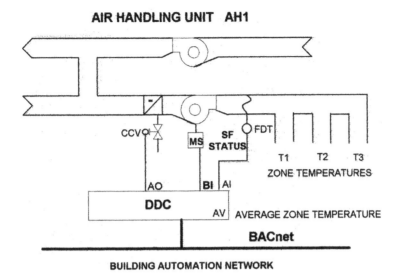

FIGURE 12.5   BI — Supply fan status.

TABLE 12.5
BACnet Binary Input Object — Supply Fan Status (SFS)

Property Identifier	Property Datatype	Property Datatype Example	Conformance Code	Comments and Examples
Object_Identifier	BACnetObjectIdentifier	X'00000001'	R	Key property, 1st binary input
Object_Name	CharacterString	"AH1SFS"	R	Use site-specific naming conventions
Object_Type	BACnetObjectType	Binary_Input	R	Description of Object type
Present_Value	BACnetBinaryPV	ACTIVE	R, W	W-when Out_Of_Service = True; see Table 12.6 for polarity relationships
Description	CharacterString	"Supply Fan Status"	O	Use site-specific naming conventions
Device_Type	CharacterString	"DP switch dry contact"	O	Use actual field device type
Status_Flags	BACnetStatusFlags	{False, False, False, False}	R	4Flags: {In_Alarm, Fault, Overriden, Out_Of_Service};Status:True (1), False (0)
Event_State	BACnetEventState	Normal	R	Determines if the object has an associated event state
Reliability	BACnetReliability	No_Fault_Detected	O	Values: {No_Fault_Detected, No_Sensor, Open_Loop, Shorted_Loop, Unreliable_Other)

TABLE 12.5 (continued)
BACnet Binary Input Object — Supply Fan Status (SFS)

Property Identifier	Property Datatype	Property Datatype Example	Conformance Code	Comments and Examples
Out_Of_Service	Boolean	False	R	True=the device is out of service; False=device in service
Polarity	BACnetPolarity	Normal	R	Normal=ACTIVE or ON;REVERSE=INACTIVE or OFF, if Out_Of_Service=FALSE
Inactive_Text	CharacterString	"Supply Fan is OFF"	O	If Inactive_Text is present, then Active_Text is also present
Active_Text	CharacterString	"Supply Fan is ON"	O	If Active_Text is present, then Inactive_Text is also present
Change_Of_State_Time	BACnetDateTime	14-Feb-1998, 10:45:30.1	O	If one of the COS or COT is present, then all of these properties shall be present
Change_Of_State_Count	Unsigned	55	O	If one of the COS or COT is present, then all of these properties shall be present; the numerical value indicates the number of PV changes since the last Change_Of_State_Count was set to zero
Change_Of_State_Count _Reset	BACnetDateTime	1-Jan-1998, 19:55:40.1	O	If one of the COS or COT is present, then all of these properties shall be present; the date and time indicates the last reset time
Elapsed_Active_Time	Unsigned	60	O	If Elapsed_Active_Time is present, then Time_Of_Active_Time_Reset shall also be present
Time_Of_Active_Time _Reset	BACnetDateTime	1-Jan-1998, 19:55:40.1	O	If Time_Of_Active_Time_Reset is present, then Elapsed_Active_Time shall also be present
Time_Delay	Unsigned	20	O, R	Defined in seconds; the PV = Alarm Value prior to reporting a To-OFF Normal or not equal to Alarm Value before To-Normal condition R=if intrinsic reporting is supported

TABLE 12.5 (continued)
BACnet Binary Input Object — Supply Fan Status (SFS)

Property Identifier	Property Datatype	*Property Datatype Example*	Conformance Code	*Comments and Examples*
Notification_Class	Unsigned	2	O, R	*BACnet defines a Notification Class object with properties for event notification over the BACnet* *R=if intrinsic reporting is supported.*
Alarm_Value	BACnetBinaryPV	*INACTIVE*	O, R	*R=if intrinsic reporting is supported*
Event_Enable	BACnetEventTransitionBits	*{True, False, True}*	O, R	*Three flags: TO-OFF NORMAL, TO-FAULT, TO-NORMAL* *R=if intrinsic reporting is supported*
Acked_Transitions	BACnetEventTransitionBits	*{True, True, True}*	O, R	*Three flags: TO-OFF NORMAL, TO-FAULT, TO-NORMAL* *R=if intrinsic reporting is supported*
Notify_Type	BACnetNotifyType	*Alarm*	O, R	*Events or Alarms* R=if intrinsic reporting is supported

TABLE 12.6
BACnet Polarity Relationships

Present Value	Polarity	Physical State of BI or BO Object	Physical State of Device
Inactive	Normal	Off or Inactive	NOT running
Active	Normal	On or Active	Running
Inactive	Reverse	ON or Active	NOT running
Active	Reverse	Off or Inactive	Running

have to be defined. Figure 12.6 and Table 12.7 illustrate and describe an example of a supply fan control (SFC) definition of an analog output object.

PRIORITY ASSIGNMENTS

The BACnet standard provides a priority table with 16 available priorities. Some priorities are already assigned; some are available for the facility and the DDC programmer to use to fit the site-specific needs and conditions. Also, interpretation of the already assigned priorities is site specific as far as what constitutes, for example, "Manual Life Safety" equipment.

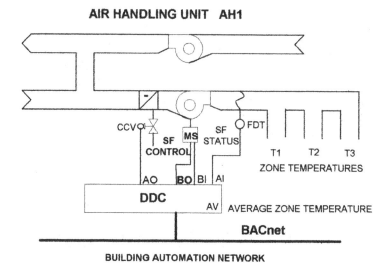

AIR HANDLING UNIT   AH1

FIGURE 12.6   AO — Supply fan control.

TABLE 12.7
BACnet Binary Output Object — Supply Fan Control (SFC)

Property Identifier	Property Datatype	Property Datatype Example	Conformance Code	Comments and Examples
Object_Identifier	BACnetObjectIdentifier	X'00000001'	R	Key property, 1st binary output
Object_Name	CharacterString	"AH1SFC"	R	Use site-specific naming conventions
Object_Type	BACnetObjectType	Binary_Output	R	Description of object type
Present_Value	BACnetBinaryPV	ACTIVE	W	See Table 12.6 for polarity relationships
Description	CharacterString	"Supply Fan Control"	O	Use site-specific naming conventions
Device_Type	CharacterString	"Interposing relay"	O	Use actual field device type
Status_Flags	BACnetStatusFlags	{False, False, False, False}	R	4Flags: {In_Alarm, Fault, Overriden, Out_Of_Service}; Status: True (1), False (0)
Event_State	BACnetEventState	Normal	R	Determines if the object has an associated event state
Reliability	BACnetReliability	No_Fault_Detected	O	Values: {No_Fault_Detected, No_Output, Open_Loop, Shorted_Loop, Unreliable_Other)
Out_Of_Service	Boolean	False	R	True=the device is out of service; False=device in service

TABLE 12.7 (continued)
BACnet Binary Output Object — Supply Fan Control (SFC)

Property Identifier	Property Datatype	Property Datatype Example	Conformance Code	Comments and Examples
Polarity	BACnetPolarity	Normal	R	Normal=ACTIVE or ON;REVERSE=INACTIVE or OFF, if Out_Of_Service=FALSE
Inactive_Text	CharacterString	"Supply Fan turned OFF"	O	If Inactive_Text is present, then Active_Text is also present
Active_Text	CharacterString	"Supply Fan turned ON"	O	If Active_Text is present, then Inactive_Text is also present
Change_Of_State_Time	BACnetDateTime	14-Feb-1998, 10:45:30.0	O	If one of the COS or COT is present, then all of these properties shall be present
Change_Of_State_Count	Unsigned	54	O	If one of the COS or COT is present, then all of these properties shall be present; the numerical value indicates the number of PV changes since the last Change_Of_State_Count was set to zero
Change_Of_State_Count _Reset	BACnetDateTime	1-Jan-1998, 19:55:40.0	O	If one of the COS or COT is present, then all of these properties shall be present; the date and time indicates the last reset time
Elapsed_Active_Time	Unsigned	59	O	If Elapsed_Active_Time is present, then Time_Of_Active_Time_Reset shall also be present
Time_Of_Active_Time _Reset	BACnetDateTime	1-Jan-1998, 19:55:40.0	O	If Time_Of_Active_Time_Reset is present, then Elapsed_Active_Time shall also be present
Minimum_Off_Time	Unsigned	180	O	In seconds
Minimum_On_Time	Unsigned	300	O	In seconds
Priority_Array	BACnetPriorityArray	Null	R	No existing command at that priority see priority Table 12.6
Relinquish_Default	BACnetBinaryPV	INACTIVE	R	The default value to be used for PV

TABLE 12.7 (continued)
BACnet Binary Output Object — Supply Fan Control (SFC)

Property Identifier	Property Datatype	Property Datatype Example	Conformance Code	Comments and Examples
Time_Delay	Unsigned	20	O, R	Defined in seconds; the PV = Alarm Value prior to reporting a To-OFF Normal or not equal to Alarm Value before To-Normal condition R=if intrinsic reporting is supported
Notification_Class	Unsigned	2	O, R	BACnet defines a notification class object with properties for event notification over the BACnet R=if intrinsic reporting is supported
Feedback_Value	BACnetBinary PV	INACTIVE	O, R	Status of feedback to generate Event
Event_Enable	BACnetEventTransitionBits	{True, False, True}	O, R	Three flags: TO-OFF NORMAL, TO-FAULT, TO-NORMAL R=if intrinsic reporting is supported
Acked_Transitions	BACnetEventTransitionBits	{True, True, True}	O, R	Three flags: TO-OFF NORMAL, TO-FAULT, TO-NORMAL R=if intrinsic reporting is supported
Notify_Type	BACnetNotifyType	Alarm	O, R	Events or Alarms. R=if intrinsic reporting is supported

BINARY VALUE OBJECT

Binary value objects are the results of calculations residing in the memory of a computer or controller connected to the BACnet-compatible network. In the past, we used to refer to them as "pseudo-points," "software points," etc. They do not have associated field hardware (i.e., relay). Such object may be, for example, a binary output to close the minimum air damper during morning warm-up, calculated from analog input objects such as outside air temperature (OAT) and return air temperature (RAT) sensors. If such objects are to be made visible to other nodes (controllers, computers, etc.) on the BACnet-compatible network, their properties have to be defined. Figure 12.7 and Table 12.8 illustrate and describe an example of a minimum air damper control (MINADC) definition of a binary value object.

MULTISTATE INPUT OBJECT

Multistate input objects are the results of calculations and/or a combination of binary or analog inputs connected to the BACnet-compatible network. Such object may be, for example, the binary

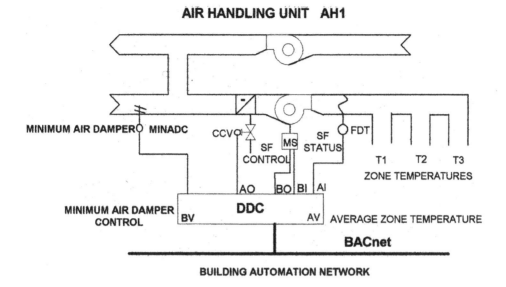

AIR HANDLING UNIT   AH1

FIGURE 12.7   BV — Minimum air damper control.

TABLE 12.8
BACnet Binary Value Object — Minimum Air Damper Control (MINADC)

Property Identifier	Property Datatype	Property Datatype Example	Conformance Code	Comments and Examples
Object_Identifier	BACnetObjectIdentifier	X'00000001'	R	Key property, 1st binary value
Object_Name	CharacterString	"AH1MINADC"	R	Use site-specific naming conventions
Object_Type	BACnetObjectType	Binary_Value	R	Description of Object type
Present_Value	BACnetBinaryPV	ACTIVE	W	See Table 12.6 for polarity relationships
Description	CharacterString	"Minimum Air Damper Control"	O	Use site-specific naming conventions
Status_Flags	BACnetStatusFlags	{False, False, False, False}	R	4Flags: {In_Alarm, Fault, Overriden, Out_Of_Service}; Status: True (1), False (0)
Event_State	BACnetEventState	Normal	R	To determine if the object has an associated event state
Reliability	BACnetReliability	No_Fault_Detected	O	Values: {No_Fault_Detected, Unreliable_Other}
Out_Of_Service	Boolean	False	R	True=PV is prevented from being locally modified
Inactive_Text	CharacterString	"Minimum Air Damper Open"	O	If Inactive_Text is present, then Active_Text is also present

TABLE 12.8 (continued)
BACnet Binary Value Object — Minimum Air Damper Control (MINADC)

Property Identifier	Property Datatype	Property Datatype Example	Conformance Code	Comments and Examples
Active_Text	CharacterString	*"Minimum Air Damper Close"*	O	*If Active_Text is present, then Inactive_Text is also present*
Change_Of_State_Time	BACnetDateTime	*14-Feb-1998, 10:45:40.0*	O	*If one of the COS or COT is present, then all of these properties shall be present*
Change_Of_State_Count	Unsigned	*60*	O	*If one of the COS or COT is present, then all of these properties shall be present; the numerical value indicates the number of PV changes since the last Change_Of_State_Count was set to zero.*
Change_Of_State_Count _Reset	BACnetDateTime	*1-Jan-1998, 18:55:40.0*	O	*If one of the COS or COT is present, then all of these properties shall be present; the date and time indicates the last reset time*
Elapsed_Active_Time	Unsigned	*79*	O	*If Elapsed_Active_Time is present, then Time_Of_Active_Time_Reset shall also be present*
Time_Of_Active_Time _Reset	BACnetDateTime	*1-Jan-1998, 18:55:40.0*	O	*If Time_Of_Active_Time_Reset is present, then Elapsed_Active_Time shall also be present*
Minimum_Off_Time	Unsigned	*120*	O	*In seconds*
Minimum_On_Time	Unsigned	*180*	O	*In seconds*
Priority_Array	BACnetPriorityArray	*Null*	R	*No existing command at that priority*
Relinquish_Default	BACnetBinaryPV	*INACTIVE*	R	*The default value to be used for PV*
Time_Delay	Unsigned	*15*	O, R	*Defined in seconds; the PV = Alarm Value prior to reporting a To-OFF Normal or not equal to Alarm Value before To-Normal condition R=if intrinsic reporting is supported*

TABLE 12.8 (continued)
BACnet Binary Value Object — Minimum Air Damper Control (MINADC)

Property Identifier	Property Datatype	Property Datatype Example	Conformance Code	Comments and Examples
Notification_Class	Unsigned	3	O, R	BACnet defines a notification class object with properties for event notification over the BACnet R=if intrinsic reporting is supported
Alarm_Value	BACnetBinary PV	INACTIVE	O, R	Status of feedback to generate Event
Event_Enable	BACnetEventTransitionBits	{True, False, True}	O, R	Three flags: TO-OFF NORMAL, TO-FAULT, TO-NORMAL R=if intrinsic reporting is supported
Acked_Transitions	BACnetEventTransitionBits	{True, True, True}	O, R	Three flags: TO-OFF NORMAL, TO-FAULT, TO-NORMAL R=if intrinsic reporting is supported
Notify_Type	BACnetNotifyType	Alarm	O, R	Events or Alarms. R=if intrinsic reporting is supported

input from a two-speed fan. If such objects are to be made visible to other nodes (controllers, computers, etc.) on the BACnet-compatible network, their properties have to be defined. Figure 12.8 and Table 12.9 illustrate and describe an example of a toilet exhaust fan status (TEFS) multistate input object.

MULTISTATE OUTPUT OBJECT

Multistate output objects represent one or more outputs to processes or outputs to controlled HVAC equipment connected to BACnet controllers on the network. Such object may be, for example, the binary output to a two-speed fan. If such objects are to be made visible to other nodes (controllers, computers, etc.) on the BACnet-compatible network, their properties have to be defined. Figure 12.9 and Table 12.10 illustrate and describe an example of a toilet exhaust fan control (TEFC) multistate output object.

LOOP OBJECT

BACnet standardizes the properties of a control (i.e., PID) loop with feedback controlling HVAC equipment; for example, a zone reheat coil connected to BACnet controllers on the network. Loop objects may simplify object definitions in some applications. However, there may be other applications where the engineers wish to use PID controllers in their vendor-specific controllers and import analog values from sensors connected to other vendor controllers. In such cases the use of analog input/output objects would be more appropriate than the loop object.

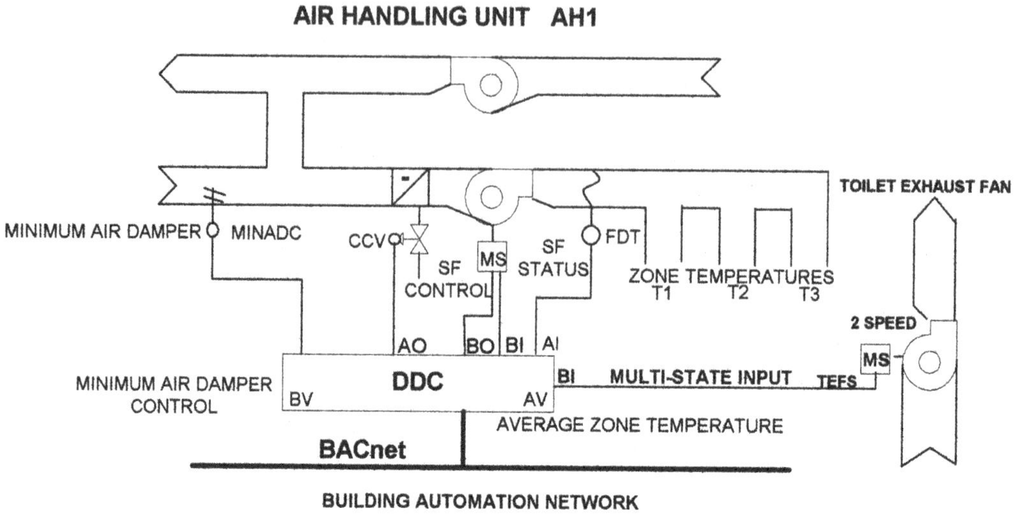

AIR HANDLING UNIT   AH1

FIGURE 12.8   Multistage input — toilet exhaust fan status.

TABLE 12.9
BACnet Multistate Input Object — Toilet Exhaust Fan Status (TEFS)

Property Identifier	Property Datatype	Property Datatype Example	Conformance Code	Comments and Examples
Object_Identifier	BACnetObjectIdentifier	X'00000001'	R	Key property, 1st multistat input
Object_Name	CharacterString	"AH1TEFS"	R	Use site-specific naming conventions
Object_Type	BACnetObjectType	Multistate_Input	R	Description of Object type
Present_Value	Unsigned	3	R, W	1= Low + High speed = OFF; 2 = Low speed = ON; 3 = High speed = ON; W, if Out_Off_Service = True
Description	CharacterString	"Toilet Exhaust Fan Status"	O	Use site-specific naming conventions
Device_Type	CharacterString	"Dry Contacts"	O	Actual field device
Status_Flags	BACnetStatusFlags	{False, False, False, False}	R	4Flags: {In_Alarm, Fault, Overriden, Out_Of_Service}; Status: True (1), False (0)
Event_State	BACnetEventState	Normal	R	To determine if the object has an associated event state
Reliability	BACnetReliability	No_Fault_Detected	O	Values: {No_Fault_Detected, No_Sensor, Over_Range, Under_Range, Open_Loop, Shorted_Loop, Unreliable_Other)

TABLE 12.9 (continued)
BACnet Multistate Input Object — Toilet Exhaust Fan Status (TEFS)

Property Identifier	Property Datatype	*Property Datatype Example*	Conformance Code	*Comments and Examples*
Out_Of_Service	Boolean	*False*	R	*True = input devices out of service*
Number_Of_States	Unsigned	*3*	R	*R, if intrinsic reporting is required*
State_Text	BACnetArray Of CharacterString	*High speed = ON*	O	*1= Low + High speed = OFF; 2 = Low speed = ON; 3 = High speed = ON*
Time_Delay	Unsigned	*10*	O, R	*Defined in seconds; the PV = Alarm Value prior to reporting a To-OFF Normal or not equal to Alarm Value before To-Normal condition R=if intrinsic reporting is supported*
Notification_Class	Unsigned	*5*	O, R	*BACnet defines a notification class object with properties for event notification over the BACnet R=if intrinsic reporting is supported*
Alarm_Values	List of Unsigned	*INACTIVE*	O, R	*Status of feedback to generate Events TO-OFFNORMAL and TO-NORMAL*
Fault_Values	List of Unsigned	*INACTIVE*	O, R	*Status of feedback to generate Events TO-FAULT and TO-NORMAL R=if intrinsic reporting is supported*
Event_Enable	BACnetEventTransitionBits	*{True, False, True}*	O, R	*Three flags: TO-OFF NORMAL, TO-FAULT, TO-NORMAL R=if intrinsic reporting is supported*
Acked_Transitions	BACnetEventTransitionBits	*{True, True, True}*	O, R	*Three flags: TO-OFF NORMAL, TO-FAULT, TO-NORMAL R=if intrinsic reporting is supported*
Notify_Type	BACnetNotifyType	*Alarm*	O, R	*Events or Alarms. R=if intrinsic reporting is supported*

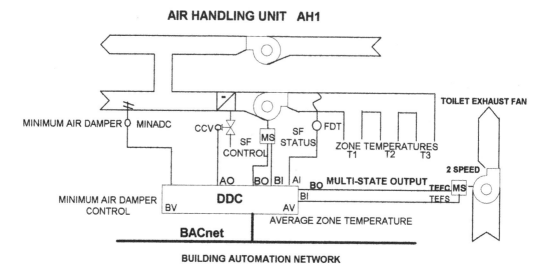

FIGURE 12.9   Multistage output object — toilet exhaust fan control.

TABLE 12.10
BACnet Multistate Output Object — Toilet Exhaust Fan Control (TEFC)

Property Identifier	Property Datatype	Property Datatype Example	Conformance Code	Comments and Examples
Object_Identifier	BACnetObjectIdentifier	X'00000001'	R	Key property, 1st multistage output
Object_Name	CharacterString	"AH1TEFC"	R	Use site-specific naming conventions
Object_Type	BACnetObjectType	Multistate_Output	R	Description of object type
Present_Value	Unsigned	3	R, W	1= Low + High speed = OFF; 2 = Low speed = ON; 3 = High speed = ON; W, if Out_Off_Service = True
Description	CharacterString	"Toilet Exhaust Fan Control"	O	Use site-specific naming conventions
Device_Type	CharacterString	"Motor Starter"	O	Actual field device
Status_Flags	BACnetStatusFlags	{False, False, False, False}	R	4Flags: {In_Alarm, Fault, Overriden, Out_Of_Service}; Status: True (1), False (0)
Event_State	BACnetEventState	Normal	R	Determines if the object has an associated event state
Reliability	BACnetReliability	No_Fault_Detected	O	Values: {No_Fault_Detected, Open_Loop, Shorted_Loop, No_Output, Unreliable_Other}
Out_Of_Service	Boolean	False	R	True if output devices is out of service

TABLE 12.10 (continued)
BACnet Multistate Output Object — Toilet Exhaust Fan Control (TEFC)

Property Identifier	Property Datatype	Property Datatype Example	Conformance Code	Comments and Examples
Number_Of_States	Unsigned	3	R	1= Low + High speed = STO; 2 = Low speed = STR; 3 = High speed = STR; R, if intrinsic reporting is required
State_Text	BACnetArray Of CharacterString	High speed = STR	O	1= Low + High speed = STO; 2 = Low speed = STR; 3 = High speed = STR
Priority_Array	BACnetPriorityArray	Null	R	No existing command at that priority
Relinquish_Default	Unsigned	INACTIVE	R	The default value to be used for PV
Time_Delay	Unsigned	10	O, R	Defined in seconds; the PV = Alarm Value prior to reporting a To-OFF Normal or not equal to Alarm Value before To-Normal condition R=if intrinsic reporting is supported
Notification_Class	Unsigned	5	O, R	BACnet defines a notification class object with properties for event notification over the BACnet R=if intrinsic reporting is supported
Feedback_Values	Unsigned	INACTIVE	O, R	Status of feedback to generate Events TO-OFFNORMAL and TO-NORMAL
Event_Enable	BACnetEventTransitionBits	{True, False, True}	O, R	Three flags: TO-OFF NORMAL, TO-FAULT, TO-NORMAL R=if intrinsic reporting is supported
Acked_Transitions	BACnetEventTransitionBits	{True, True, True}	O, R	Three flags: TO-OFF NORMAL, TO-FAULT, TO-NORMAL R=if intrinsic reporting is supported
Notify_Type	BACnetNotifyType	Alarm	O, R	Events or Alarms R=if intrinsic reporting is supported

If loop objects are to be made visible to other nodes (controllers, computers, etc.) on the BACnet-compatible network, their properties have to be defined. Figure 12.10 and Table 12.11 illustrate and describe an example of a zone reheat coil loop (ZRHL) loop object.

The PID loop control definition of the zone reheat is also shown in Figure 12.11.

DEVICE OBJECT

BACnet standardizes the properties of BACnet devices (i.e., controllers) on the network. If such objects are to be made visible to other nodes (controllers, computers, VT terminal, etc.) on the BACnet-compatible network, their properties have to be defined. Figure 12.12 and Table 12.12

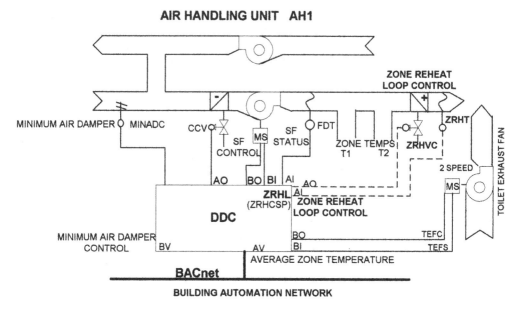

FIGURE 12.10   Loop object — zone reheat coil control.

TABLE 12.11
BACnet Loop Object — Zone Reheat Coil Loop Control (ZRHLC)

Property Identifier	Property Datatype	Property Datatype Example	Conformance Code	Comments and Examples
Object_Identifier	BACnetObjectIdentifier	X'00000001'	R	Key property, 1st loop object
Object_Name	CharacterString	"AH1ZRHL"	R	Use site-specific naming conventions
Object_Type	BACnetObjectType	Loop	R	Description of object type
Present_Value	REAL	42.5	R	Current output of the loop algorithm
Description	CharacterString	"Zone reheat coil PI control loop"	O	Use site-specific naming conventions
Status_Flags	BACnetStatusFlags	{False, False, False, False}	R	4Flags: {In_Alarm, Fault, Overriden, Out_Of_Service}; Status: True (1), False (0)
Event_State	BACnetEventState	Normal	R	Determines if the object has an associated event state
Reliability	BACnetReliability	No_Fault_Detected	O	Values: {No_Fault_Detected, Open_Loop, Unreliable_Other)
Out_Of_Service	Boolean	False	R	True-output devices out of service

TABLE 12.11 (continued)
BACnet Loop Object — Zone Reheat Coil Loop Control (ZRHLC)

Property Identifier	Property Datatype	Property Datatype Example	Conformance Code	Comments and Examples
Update_Interval	Unsigned	20	O	Output update by loop algorithms in milliseconds
Output_Units	BACnetEngineeringUnits	%	R	PV value
Manipulated_Variable _Reference	BACnetObjectProperty Reference	X'00000002	R	Analog Output, Instance 2 =ZRHCVC,
Controlled_Variable_ Reference	BACnetObjectProperty Reference	X'00000002	R	Analog Input, Instance 2=ZRHT,
Controlled_Variable_Value	REAL	65.8	R	PV of ZRHT=65.8degF
Controlled_Variable_Units	BACnetEngineeringUnits	degF	R	BACnet-defined engineering unit
Setpoint_Reference	BACnetSetpointReference	X'00000002	R	Analog_Value, Instance 2-ZRHCSP
Setpoint	REAL	65	R	PV of ZRHCSP in degF
Action	BacnetAction	DIRECT	R	Direct or REVERSE acting valve
Proportional_Constant	REAL	.4	O, R	R if Proportional_Constant_ Unit is present
Proportional_Constant_Unit	BACnetEngineeringUnit	% per dgF	O, R	R, if Proportional_Constant is present
Integral_Constant	REAL	.3	O, R	R if Integral_Constant_ Unit is present
Integral_Constant_Unit	BACnetEngineeringUnit	Per-minute	O, R	R, if Integral_Constant is present
Derivative_Constant	REAL	.0	O, R	R if Derivative_Constant_ Unit is present
Derivative_Constant_Unit	BACnetEngineeringUnit	No-units	O, R	R, if Derivative_Constant is present
Bias	REAL	6.0	O	In % of output unit
Maximum_Output	REAL	100	O	100% = fully opened
Minimum_Output	REAL	0	O	0% = fully closed
Priority_For_Writing	Unsigned	14	R	One of 16 priorities assigned
COV_Increment	REAL	.5	O, R	Change of PV to cause COVNotification; R, if COV reporting is supported

TABLE 12.11 (continued)
BACnet Loop Object — Zone Reheat Coil Loop Control (ZRHLC)

Property Identifier	Property Datatype	Property Datatype Example	Conformance Code	Comments and Examples
Time_Delay	Unsigned	*4*	O, R	*R, if intrinsic reporting is supported*
Notification_Class	Unsigned	*1*	O, R	*BACnet defines a notification class object with properties for event notification over the BACnet R=if intrinsic reporting is supported*
Error_Limit	REAL	*5.0*	O, R	*The difference between ZRHTSP and ZRHT must be greater than %degF to generate TO-OFFNORMAL event R=if intrinsic reporting is supported*
Event_Enable	BACnetEventTransitionBits	*{True, False, True}*	O, R	*Three flags: TO-OFF NORMAL, TO-FAULT, TO-NORMAL R=if intrinsic reporting is supported*
Acked_Transitions	BACnetEventTransitionBits	*{True, True, True}*	O, R	*Three flags: TO-OFF NORMAL, TO-FAULT, TO-NORMAL R=if intrinsic reporting is supported*
Notify_Type	BACnetNotifyType	*Alarm*	O, R	*Events or Alarms R=if intrinsic reporting is supported*

illustrate and describe an example of a definition of a unique BACnet controller residing on the BACnet network.

The above examples illustrate the BACnet object definitions and their properties with examples as they relate to building automation systems.

OTHER BACNET OBJECTS

BACnet defines additional objects typical in the building automation industry. The following contains the short description of the additional BACnet objects without illustrations and tabulation of their properties:

- **Event Enrollment Object** — Event enrollment objects assure transmission of change of value of any property of a BACnet object to one or more controllers or computers connected to the BACnet network. The event types specified in the BACnet are

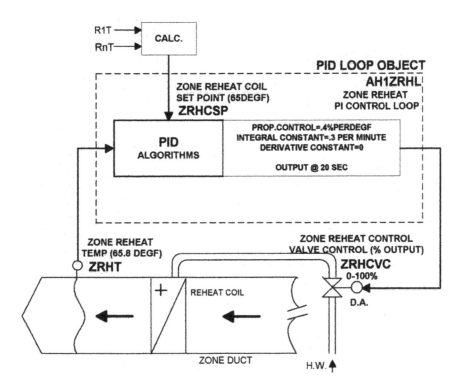

FIGURE 12.11 PID loop control.

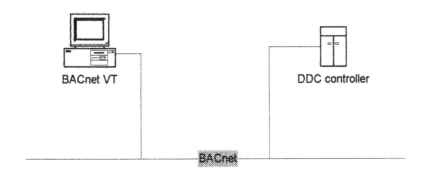

FIGURE 12.12 BACnet devices on a network.

TABLE 12.12
BACnet Device Object — DDC Controller

Property Identifier	Property Datatype	Property Datatype Example	Conformance Code	Comments and Examples
Object_Identifier	BACnetObjectIdentifier	X'00000001'	R	Key property, 1st device object
Object_Name	CharacterString	"AH1DDC"	R	Use site-specific naming conventions
Object_Type	BACnetObjectType	DEVICE	R	Description of object type

TABLE 12.12 (continued)
BACnet Device Object — DDC Controller

Property Identifier	Property Datatype	Property Datatype Example	Conformance Code	Comments and Examples
System_Status	BACnetDeviceStatus	OPERATIONAL	R	OPERATIONAL, OPERATIONAL_READ_ONLY, DOWNLOAD_REQUIRED, DOWNLOAD_IN_PROGRESS, NON_OPERATIONAL
Vendor_Name	CharacterString	"XYZ CONTROLS COMPANY"	R	Use vendor-specific name
Vendor_Identifier	Unsigned16	1234	R	Use site-specific conventions
Model_Name	CharacterString	1010	R	Use vendor-specific conventions
Firmware_Revision	CharacterString	1.1	R	Use vendor-specific conventions
Application_Software _Version	CharacterString	"SW 1.1 VJB May 1,1998"	R	Use vendor-specific conventions
Location	CharacterString	"MR101"	O	Use site-specific naming conventions
Description	CharacterString	"AH1 DDC Controller"	O	Use site-specific naming conventions
Protocol_Version	Unsigned	1	R	Initial BACnet release
Protocol_Conformance _Class	Unsigned(1-6)	6	R	Range 1–6, see BACnet classes
Protocol_Service _Supported	BACnetServiceSupported		R	Indicates which protocol services are supported by the controller
Protocol_Object_Type _Supported	BACnetObjectType Supported		R	Indicates which standard object types are supported by the controller
Object_List	BACnetARRAYof BACnetObjectIdentifier		R	List of objects in the controller accessible via BACnet
Max_APDU_Length _Accepted	Unsigned	400	R	Constrained by data link technology
Segmentation_Supported	BacnetSegmentation	SEGMENTED_BOTH	R	SEGMENTED_BOTH, SEGMENTED_TRANSMIT, SEGMENTED_RECEIVE, NO_SEGMENTATION
VT_Classes_Supportednt	ListofBACnetVTClass	VT 100,DEFAULT _TERMINAL	O	DEFAULT_TERMINAL is the minimum to be supported
Active_VT_Sessions	ListofBACnetVTSession		O	Consists of Local VT Session ID, Remote VT Session ID, and Remote VT Address

TABLE 12.12 (continued)
BACnet Device Object — DDC Controller

Property Identifier	Property Datatype	Property Datatype Example	Conformance Code	Comments and Examples
Local_Time	Time	"01:23:44.56"	O	
Local_Date	Date	"May 1, 1998	O	
UTC_Offset	INTEGER	4	O	Offset of local time from UTC
Daylight_Savings_Status	BOOLEAN	TRUE	O	
APDU_Segment_Timeout	Unsigned	2000	O	In milliseconds
APDU_Timeout	Unsigned	60,000	R	In milliseconds; 60,000 is the default value
Number_Of_APDU_Retries	Unsigned	3	R	
List_Of_Session_Keys	ListOfSessionBACnetKey		O	Assigned by a key server
Time_Synchronization _Recipients	ListOfBACnetRecepient		O	The controller will send a time synchronization request to devices listed
Max_Master	Unsigned	13	O	Applies to MS/TP networks
Max_Info_Frames	Unsigned	3	O	Applies to MS/TP networks; the max. number of frames the controller may send before passing a token
Device_Address_Binding	ListOfBACnetAddress Binding		R	A list of associated BACnet_Object_Identifiers, and BACnet_Addresses

{CHANGE_OF_BITSTRING, CHANGE_OF_STATE, CHANGE_OF_VALUE, COMMAND_FAILURE, FLOATING_LIMIT, OUT_OF_RANGE}. The state of events (i.e., NORMAL, OFFNORMAL, HIGH_LIMIT, LOW_LIMIT) as well as their associated parameters (i.e., Time_Delay, Bitmask, values, references, etc.), are also specified in the standard. Further, the recipient(s) of the event, process identifier, priorities, and other properties are also defined.

- **Notification Class** — Provides prioritization of handling TO_OFFNORMAL, TO_FAULT, TO_NORMAL events as per their assigned priorities (0, the highest; 255, the lowest priority). The notification class object also has "required properties" for event acknowledgment and a list of recipients to which the notification will be sent at specified days of the week, time, etc.
- **Command Object** — This object type is more than just sending a simple ON/OFF command to a device. It could involve a complex set of conditions and each state could involve writing to properties of numerous objects that require to be changed. For example, a command object "Laboratory Control" can initiate a sequence of actions based on a defined present value. A present value of, for example, an "Occupied mode" can trigger a command to a BACnet controller in the basement controlling the hot water exchanger to increase the set-point of the hot water supply temperature for "Occupied mode" of operation.

Another command can be issued to a BACnet HVAC controller to increase the fan speed to meet the required ventilation rate and increase the set-point of the fan discharge air temperature. Yet another command can be sent to fume hood controllers to increase the face velocity of the air in the fume hood, etc. Similar commands can be issued to numerous controllers for "Standby" (warm-up, cool-down), or "Unoccupied" modes of operations. These actions are to be defined in the "Action" property of the command object.

- **Group Object** — Group objects defined on the BACnet network can be based on their common characteristics. For example, room temperatures of an area of a college campus (i.e., student dorms) can be grouped together. Their properties have to be defined in the "List_Of_Group_Members" and in the "Present_Value" property of the Group Object. Any combination of object types can be included in the same Group Object (i.e., in the case of the student dorm temperatures, the respective hot water supply temperatures of the heating systems can be grouped for each dorm).

- **File Object** — This object provides access to files defined in the BACnet and also in vendor-specific devices. BACnet File Access Services define the methods of file access, read, write operations, and other particular items pertaining to file access. The DDC system in a building may collect data and store it in a history file for trending or report generation. The desired data can be accessed from another BACnet device by defining the properties of the File Object.

- **Program Object** — This provides a view of selected parameters of an application program running on a local controller. For example, the property "Program_State" reflects the state of the applications program by values IDLE (not being executed), LOADING (being loaded), RUNNING (currently being executed), WAITING (for an external value), HALTED (due to an error), UNLOADING (to terminate the process). Other properties, such as "Program_Change," "Reason_For_Halt," and "Description_Of_Halt" provide the desired information on the object to BACnet devices on the network.

- **Schedule Object** — Relates to definition of schedules used in building automation systems connected to the BACnet network. Each BAS system has schedule "schedules." However, they are different from vendor to vendor. The BACnet schedule object defines periodical schedules, but also includes definition of exceptions (i.e., a classroom schedule for an academic year with an exception for exam dates).

- **Calendar Object** — This object describes, say, a list "Holidays" in an academic year for a school district. If the Present_Value of the Calendar Object is TRUE, the associated systems are going to be reset into, say, an UNOCCUPIED mode of operation.

FROM APPLICATION PROGRAMS TO BACnet APPLICATION LAYERS

Data from BAS application programs to be shared by other BACnet-compatible controllers or computers over the network have to be interfaced to the BACnet application layer via an application program interface (API) (Figure 12.13). Application programs are not affected by the BACnet protocol and are specific to individual vendors' building automation systems. This assures the proprietary nature of individual systems and their applications programs. Individual control vendors utilizing BACnet communications protocol have to write API subroutines to interface their application programs to other BACnet-compatible devices on the network. It means that they have to use BACnet API services for transporting packets of information from one system to another.

The BACnet application layer is composed of two components: **BACnet User Element** (UE), and **BACnet Application Service Element** (ASE).

Information from one controller or computer to another, via the BACnet protocol, is sent from the application program to the application layer of the protocol via an API. Network control, logical link control (LLC), and media access control (MAC) related parameters from API (i.e., identity address, protocol control information) go directly to the protocol's network, data link, and physical

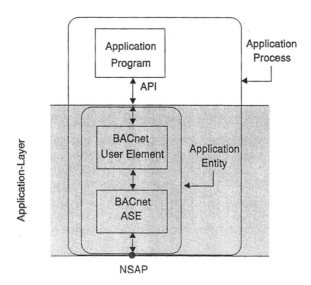

FIGURE 12.13   Block diagram of a BACnet application process. (From BACnet, a Data Communication Protocol for Building Automation and Control Networks, ANSI/ASHRAE Standard 135-1995. With permission.)

layers. Application service-related parameters (primitives) go from the API to UE and ASE, generating application protocol data unit (APDU). The protocol data unit (PDU) is the data element of a network service primitive passed through the stacks of network layers.

BACnet UE provides several functions, such as supporting API, maintaining information on the context of the transaction, maintaining time-out counters required for retrying transmissions, and mapping over activities into the BACnet objects.

The ASE provides the following application services: alarm and event services, file access services, object access services, remote device management services, and virtual terminal services.

The four service primitives passed through the stacks of network layers are: **request**, **indication**, **response**, and **confirm**, defined for BACnet PDUs.

There are four interface control information (ICI) parameters: **Destination_Address** (DA), **Source_Address** (SA), **Network_Priority** (NP), and **Data_Expecting_Reply** (DER), for messages to MS/TP data link layer, are applicable for various service primitives as per the Table 12.13.

For each transaction, the requesting and responding BACnet controller or computer creates a transaction state machine (TSM) when the transaction begins. Each transaction is identified by the client BACnet address requesting transaction, the server BACnet address responding to the client, and by the "Invoke ID." For information received from the network layer, the PDU type, source, and destination address will be utilized to determine the identity of the TSM to which the PDU will be passed. For information passed from the application program, the request type, source, and destination are to be utilized (Table 12.14).

BACnet standard 135-1995 provides time sequence diagrams for service primitives for normal unconfirmed service, abnormal unconfirmed service, normal confirmed service, abnormal confirmed service, and abnormal service request or response (protocol error).

The BACnet diagrams provide graphical illustration of time sequence for the above service primitives. Let us illustrate a NORMAL CONFIRMED SERVICE with their ICI (Figure 12.14). For more information refer to the relevant sections of the standard.

THE BACnet NETWORK LAYER

The network layer provides means to transfer messages to a single BACnet device on a network, such as a DDC controller or PC, broadcast to another BACnet network, or broadcast globally, to

TABLE 12.13
Applicability of ICI Parameters

Service Primitive	DA	SA	NP	DER
CONF_SERV.request	Yes	No	Yes	Yes
CONF_SERV.indication	Yes	Yes	Yes	No
CONF_SERV.response	Yes	No	Yes	No
CONF_SERV.confirm	Yes	Yes	Yes	No
UNCONF_SERV.request	Yes	No	Yes	Yes
UNCONF_SERV.indication	Yes	Yes	Yes	No
ERROR.request	Yes	No	Yes	No
ERROR.indication	Yes	Yes	Yes	No
REJECT.request	Yes	No	Yes	No
REJECT.indication	Yes	Yes	Yes	No
SEGMENT_ACK.request	Yes	No	Yes	No
SEGMENT_ACK.indication	Yes	Yes	Yes	No
ABORT.request	Yes	No	Yes	No
ABORT.indication	Yes	Yes	Yes	No

Source: BACnet, a Data Communication Protocol for Building Auto-
mation and Control Networks, ANSI/ASHRAE Standard 135-1995.
With permission.

TABLE 12.14
Two Classes of BACnet APDUs

APDUs Sent by a Client (requestor)	APDUs Sent by a Server (respondent)
Unconfirmed Request	Simple ACK
Confirmed Request	Complex ACK
	Error
	Reject
Segment ACK w/server = FALSE	Segment ACK w/server = TRUE
Abort w/server = FALSE	Abort w/server = TRUE

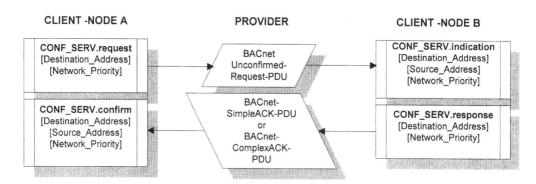

FIGURE 12.14   Normal confirmed service with interface control information.

TABLE 12.15
Maximum NPDU Length for Different Networks

Data Link	Maximum NPDU Length
Ethernet (ISO 8802-3)	1497 octets
ARCNET	501 octets
MS/TP	501 octets
Point-to-point	501 octets
LonTalk	228 octets

Source: BACnet, a Data Communication Protocol for Building Automation and Control Networks, ANSI/ASHRAE Standard 135-1995. With permission.

all or selected nodes (controllers and PCs) residing on the network. Not only controllers and PCs but also BACnet routers interconnecting two otherwise incompatible LANs, such as Ethernet and ARCNET, are using network layers to build their routing tables.

The network layer uses UNITDATA request and indication service to transfer messages to and from the application layer.

BACnet has limitations as far as maximum length of NPDU messages dependent on the data link technology; longer messages must be separated.

The standard provides detailed specifications of the network layer services, network layer PDU structure, messages for multiple recipients, network layer protocol messages, network layer procedures, BACnet routers, and point-to-point half routers, which are of great value to network and interface designers. For detailed information on the network layer, refer to the BACnet Standard SPC 135-1995.

DATA LINK/PHYSICAL LAYERS

Data link and physical layers provide interface to the physical media. Depending on the network, BACnet provides Ethernet, ARCNET, Serial, and LonTalk interfaces.

LLC uses ISO 8802-2 data link service units. The BACnet link service data unit (LSDU) consists of a BACnet network protocol data unit (NPDU). Both BACnet and Ethernet LAN (ISO 8802-3) have to conform to LLC Type 1 — Unacknowledged Connectionless-Mode service. LLC parameters are conveyed using the following DL-UNIDATA primitives:

- Source address
- Destination address
- Data
- Priority parameters

The source and destination addresses must have the MAC address and the link service access point (LSAP). *BACnet accepts all physical media recommended by ISO 8802-3 Ethernet LAN technology.*

ARCNET LAN technology used in BACnet networks must conform to the ISO 8802-2 type 1 LLC. The source and destination addresses must have the MAC address, LSAP, and a system code (SC). *BACnet accepts all physical media recommended by the ARCNET standard.*

Master-slave/token passing (MS/TP) data link protocol provides services for EIA-485 physical layer, using the following hardware:

- Universal Asynchronous Receiver/Transmitter (UART) capable of transmitting and receiving 8 data bits with 1 stop bit and no parity

- EIA-485 transceiver with disabled driver
- A timer with resolution less than 5 msec.

The connecting cable should be an 18 AWG conductor shielded twisted pair cable with characteristic impedance of 100 to 130 Ω. The maximum recommended distance is 4000 ft; the maximum number of nodes per segment is 32. Since the RS-485 technology is constantly evolving, one should be aware with the latest development and updates of the RS-485 standard (Figure 12.15).

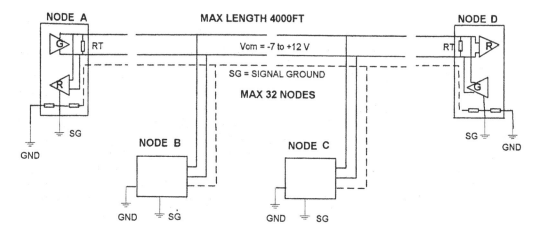

FIGURE 12.15   EIA-485 network with three different nodes.

BACnet devices may communicate using *point-to-point (PTP) communications and EIA-232*, bus-level, or line-driver interfaces. PTP connection is capable of full duplex communications between half routers connecting two networks. BACnet describes primitives and parameters needed to manage the connection and termination phases for the PTP data link layer.

BACnet provides interface to LonTalk LAN technology by passing LSDUs to LonTalk devices that confirm to ISO 8802-2 LLC Class I, type 1 LLC ACnet DL-UNITDATA primitive, consisting of source address, destination address, and priority is mapped over to the application layer interface of the LonTalk msg_send and msg_rcv request primitive. LPDUs can not be longer than 228 octets. The physical media used must meet the LonMark interoperability requirements.

APPLICATION SERVICES

Protocol application services distinguish protocols from each other and they distinguish themselves from communication drivers. The more complex the protocol, the more application services are provided by that protocol. Complex protocols distribute more information to the connected nodes on the network. Complexity of the protocol services comes at a cost. The more services the protocol provides, the higher will be the cost of the protocol, its operation, and upkeep.

The most utilized application services of the BACnet are listed below.

ALARM AND EVENT SERVICES

These services relate to change of state (COS) for *binary*, or change of value (COV) for *analog* properties of BACnet objects transmitted over the network. *Events* (nonalarms) are transmitted from one controller to another or from one computer to another, and are further utilized in vendor-specific applications or supervisory programs. *Alarm events* are like events, but they may require optional acknowledgment from an operator.

Table 12.16 lists the object types, the criteria for change of value (COV), and their reported properties.

BACnet intrinsic reporting provides alarm and event notification generated by one or more devices to one or more destinations for event transitions: TO-OFFNORMAL, TO-FAULT, TO-NORMAL.

Table 12.17 represents the object types, criteria, event types, and notification parameters for intrinsic reporting and their values returned in notification, and Table 12.18 is for six event-type algorithms and their parameters.

TABLE 12.16
Object Types and Criteria for COV and Their Properties

Object Type	Criteria	Reported Properties
Analog Input Analog Output Analog Value	Present_Value change by COV_Increment Change of Status_Flag	Present_Value Status_Flag
Binary Input Binary Output Binary Value Multistate Input Multistate Output	Change of Present_Value Change of Status_Flag	Present_Value Status_Flag
Loop	Present_Value change by COV_Increment	Present_Value Status_Flag Setpoint Controlled_Variable

TABLE 12.17
Notification Parameters for Intrinsic Reporting

Object Type	Criteria	Event Type	Notification Parameters	Referenced Object Properties
Binary Input Binary Value Multistate Input	PV changes to a new state for > than Time_Delay, and Event_Enable	CHANGE_OF_STATE	New_State Status_Flag	Present_Value Status_Flag
Analog Input Analog Output Analog Value	PV exceeds range between HI and LO_Limits for greater than Time_Delay, and Event_Enable, and PV returns to HI/LO Limit range ± Deadband for greater than Time_Delay, and Event_Enable and Limit_Enable	OUT_OF_RANGE	Exceeding_Value Status_Flag Deadband Exceeded_Limit Exceeded_Limit	Present_Value Status_Flag Deadband Low_limit or High_limit
Binary Output Multistate Output	PV is not equal to Feedback_Value for > than Time_Delay AND Event_Enable	COMMAND_FAILURE	Commanded_Value Status_Flag Feedback_Value	Present_Value Status_Flag Feedback_Value

TABLE 12.17 (continued)
Notification Parameters for Intrinsic Reporting

Object Type	Criteria	Event Type	Notification Parameters	Referenced Object Properties
Loop	If Error_Limit > than absolute difference between Setpoint and Controlled_Variable Value for > than Time_Delay and Event_Enable	FLOATING_LIMIT	Referenced_Value Status_Flag Setpoint_Value Error_Limit	Controlled_Variable_ Status_Flag Setpoint Error_Limit

TABLE 12.18
Event Types and Their Parameters

Event Type	Notification Parameters	Description
CHANGE_OF_BITSTRING	Referenced_Bitstring Status_Flags	The new value of the property The status flag of an object
CHANGE_OF_STATE	New_State Status_Flags	The new value of the property The status flag of an object
CHANGE_OF_VALUE	New_Value Status_Flags	The new value of the property The status flag of an object
COMMAND_FAILURE	Command_Value Status_Flags Feedback_Value	The value of commanded property The status flag of an object The value different from the Command_Value
FLOATING_LIMITS	Referenced_Value Status_Flags Setpoint_Value Error_Limit	The value of the property The status flag of an object The value of the setpoint The difference that was exceeded
OUT_OF_RANGE	Exceeding_Value Status_Flags Deadband Exceeded_Limits	The value exceeding the limit The status flag of an object The limit's deadband The exceeded limit

File Access Services

These function to access vendor files defined in a BACnet file object. During a file access (atomic read or write), no other read or write access is allowed to the same file (the device itself is left with the issues of synchronization — as a local matter):

- AtomicReadFile Service — to perform open-read-close operation
- AtomicWriteFile Service — to perform open-write-close operation

Object Access Services

BACnet describes nine application services to access and manipulate properties of defined BACnet objects. The majority of the BACnet communication traffic involves the following services:

- AddListElement Service — for example, there is a list of elements, such as room temperatures displayed on a floor plan, to which another element (another room temperature) is added. An object property is added to already-existing properties on the list.
- RemoveListElement Service — to remove elements from the property of a list object, for example, to remove a description from one of the properties on a defined list.
- CreateObject Service — to create standard or vendor-specific objects such as, for example, a file containing history data for trending.
- DeleteObject Service — to delete an existing object. However, BACnet allows creating groups that are not deletable. For example, in a multibuilding application, outside air temperatures (OAT) may be used to compute an average OAT. All OATs can be defined in a group as not deletable to avoid accidental deletion of one of the OATs, which would upset the calculation.
- ReadProperty Service — used to request the value of a property, for example, to read the present value of an OAT sensor connected to one of the BACnet controllers with valid ID.
- ReadPropertyConditional Service — to request object IDs and values of specified properties. The service can be used in several ways. For example, to request Object_Identifiers of a group of analog inputs, let's say, using a "wild card," or to read their conditions (i.e., room temperatures greater than 75°F), or to list all analog inputs with an unreliable status, etc.
- ReadPropertyMultiple Service — can provide information to the operator in several ways. For example, to request more than one value of a specified BACnet object (such as the temperature reading and reliability condition of a room temperature sensor), or to access the present values of a group of objects (such as chill water supply temperature, return temperature, supply pressure, return pressure, and flow).
- WriteProperty Service — to modify values of a property of a BACnet object; for example, to change a set-point value of an analog object over the BACnet.
- WritePropertyMultiple Service — to modify values of multiple properties or BACnet objects; for example, to change set-points of all zone discharge air temperatures of an air-handling unit serving a building.

REMOTE DEVICE MANAGEMENT SERVICES

These provide the following services:

- DeviceCommunicationControl — enables or disables communication to a BACnet device for a predetermined time when the device is being serviced.
- ConfirmedPrivateTransfer — to request information on a vendor ID, service, service parameters, etc., and receive confirmation from that BACnet device.
- UnconfirmedPrivateTransfer — the same as above, but no response is expected.
- ReinitializeDevice — used to reinitialize a BACnet device (warm start) online. The service will return a "Result(+)" if the password was valid and the device reinitialized; "Result (–)" if the attempt was unsuccessful.
- ConfirmedTextMessage — a message can be sent out to a BACnet device with a defined priority (i.e., "Replace all filters on AHU-1"). Successful reception of the message will result in return of "Result(+)."
- UnconfirmedTextMessage — the same as above, but no return is expected.
- TimeSynchronization — notifies all remote devices on a current date and time.
- Who-Has and I-Have — for example, an attempt to locate (Who-Has) "Bldg1 OAT" will result in a reply (I-Have) identifying the device, object, and object name of the device containing that information (i.e., the DDC controller in building 1).

- Who-Is and I-Am — a service used, for example, to establish a network address of a BACnet device or to find out the parameters of a BACnet device.

VIRTUAL TERMINAL SERVICE

The virtual terminal service facilitates bidirectional communication between client and server devices using BACnet protocols. Successful communication will result in "Result(+)," meaning that a session was established. This should be followed by an operator log-on. Successful completion of the operation will result in "Result(+)" indication.

ERROR, REJECT, AND ABORT CLASSES AND CODES

Table 12.19 was compiled to provide condensed information on the BACnet codes and classes to facilities engineers and manager. These are the BACnet protocol-related errors divided into classes, and further into codes:

TABLE 12.19
BACnet Error, Reject, and Abort Codes and Classes

Error Class	Error Codes
Device	DEVICE_BUSY, CONFIGURATION_IN_PROGRESS, OPERATIONAL_PROBLEM, OTHER
Object	DYNAMIC_CREATION_NOT_SUPPORTED, NO_OBJECTS_OF_SPECIFIED_TYPE, OBJECT_DELETION_NOT_PERMITTED, OBJECT_IDENTIFIER_ALREADY_EXISTS, READ_ACCESS_DENIED, UNKNOWN_OBJECT UNSUPPORTED_OBJECT_TYPE, OTHER
Property	CHARACTER_SET_NOT_SUPPORTED, INCONSISTENT_SELECTION_CRITERION INVALID_ARRAY_INDEX INVALID_DATA_TYPE READ_ACCESS_DENIED UNKNOWN_PROPERTY WRITE_ACCESS_DENIED VALUE_OUT_OF_RANGE, OTHER
Resources	NO_SPACE_FOR_OBJECT, NO_SPACE_TO_ADD_LIST_ELEMENT, NO_SPACE_TO_WRITE_PROPERTY, OTHER
Security	AUTHENTICATION_FAILED, CHARACTER_SET_NOT_SUPPORTED, INCOMPATIBLE_SECURITY_LEVELS, INVALID_OPERATOR_NAME, KEY_GENERATION_ERROR, PASSWORD_FAILURE, SECURITY_NOT_SUPPORTED, TIMEOUT, OTHER

TABLE 12.19 (continued)
BACnet Error, Reject, and Abort Codes and Classes

Error Class	Error Codes
Services	CHARACTER_SET_NOT_SUPPORTED, FILE_ACCESS_DENIED, INCONSISTENT_PARAMETERS, INVALID_FILE_ACCESS_METHOD, INVALID_FILE_START_POSITION, INVALID_PARAMETER_DATA_TYPE, INVALID_TIME_STAMP, PROPERTY_IS_NOT_A_LIST, MISSING_REQUIRED_PARAMETER, SERVICE_REQUEST_DENIED, OTHER
Virtual terminal	UNKNOWN_VT_CLASS, UNKNOWN_VT_SESSION, NO_VT_SESSIONS_AVAILABLE, VT_SESSION_ALREADY_CLOSED, VT_SESSION_TERMINATION_FAILURE, OTHER
Reject reason	BUFFER_OVERFLOW
Rejection of confirmed request PDUs	INCONSISTENT_PARAMETERS, INVALID_PARAMETER_DATA_TYPE, INVALID_TAG, MISSING_REQUIRED_PARAMETER, PARAMETER_OUT_OF_RANGE, TOO_MANY_ARGUMENTS, UNDEFINED_ENUMERATION, UNRECOGNIZED_SERVICE, OTHER
Abort reasons	BUFFER_OVERFLOW, INVALID_APDU_IN_THIS_STATE, PREEMPTED_BY_HIGHER_PRIORITY_TASK, SEGMENTATION_NOT_SUPPORTED, OTHER

COMMAND PRIORITIZATION

Command prioritization is well known and has been used in BAS for the last 2 decades. Prioritization determines which command execution takes precedence. For example, large motors of an AHU were commanded ON by the START/STOP program. However, the same AHU is eligible for load shedding if the demand-limiting program requires load shedding. Since the two programs may run on different systems connected to the BACnet, the two commands to the same motor starter have to be prioritized. Priority levels are assigned to each of the commanding entities. BACnet defines the commandable properties of the following objects: analog output, binary output, multistate output, analog value, and binary value.

Since any of these objects can be commanded by several means (application programs or manually), BACnet defines priority levels 1 (highest) through 16 (lowest) for WriteProperty and WritePropertyMultiple service requests which can issue commands.

Priority level 1 is assigned to *Manual-Life Safety*
Priority level 2 is assigned to *Automatic-Life Safety*
Priority levels 3 and 4 are not assigned
Priority level 5 is assigned for *Critical Equipment Control*
Priority level 6 is assigned for *Minimum On/Off*
Priority level 7 is not assigned
Priority level 8 is assigned for *Manual Operator*
Priority levels 9 through 16 are not assigned.

ENCODING BACnet PROTOCOL DATA UNITS (PDUS)

BACnet has selected ISO Standard 8824 — Specification of Abstract Syntax Notation One — ASN.(1) for data encoding. Data encoding is a specialized task for developers of protocol interfaces and is outside of the responsibilities of applications engineering. For further information refer to Standard 135-1995.

BACnet CONFORMANCE CLASSIFICATION

Since not all controllers and OWS must have the full functionality of the BACnet protocol in the building automation systems hierarchy, BACnet defines six conformance classes.

The following tables, 12.20 through 12.24, are reprinted from the ASHRAE standard for information only. Facilities engineers and managers should refer to the standard for more details and to the organization providing BACnet protocol conformance testing for more information. The main reasons the tables are reprinted in this book are to raise the users' awareness of these conformance classes and to point to the fact that not all BACnet devices conform to all of the classes or have the same conform class. Therefore, users selecting a BACnet-compatible system should be aware of the conformance issues, understand their own needs, and question the proposed systems manufacturers on the conform class they support.

The column *Init* means that the device shall be able to initiate the service request
The column *Exec* means that the device shall be able to execute the service request

Conformance Class 6 will support all requirements of Class 5, and in addition will support the following functional groups:

- Clock Function
- Hand-Held Work Station Functional Group
- Personal Computer Workstation Functional Group
- Event Initiation Functional Group
- Event Response Functional Group
- COV Event Initiation Functional Group
- COV Event Response Functional Group
- Files Functional Group
- Reinitialize Functional Group
- Virtual Operator Interface Group
- Virtual Terminal Functional Group
- Device Communications Functional Group
- Time Master Functional Group
- BACnetRouter

TABLE 12.20
Conformance Class 1

Application Service	Init	Exec	Objects
ReadProperty		×	Device

TABLE 12.21
Conformance Class 2

Application Service	Init	Exec
WriteProperty		×

TABLE 12.22
Conformance Class 3

Application Service	Init	Exec
I-Am	×	
I-Have	×	
ReadPropertyMultiple		×
WritePropertyMultiple		×
Who-Has		×
Who-Is		×

TABLE 12.23
Conformance Class 4

Application Service	Init	Exec
AddListElement	×	×
RemoveListElement	×	×
ReadProperty	×	
ReadPropertyMultiple	×	
WriteProperty	×	
WritePropertyMultiple	×	

TABLE 12.24
Conformance Class 5

Application Service	Init	Exec
CreateObject		×
DeleteObject		×
ReadPropertyConditional		×
Who-Has	×	
Who-Is	×	

Source: Tables 12.20 to 12.24, *BACnet, a Data Communication Protocol for Building Automation and Control Networks*, ANSI/ASHRAE Standard 135-1995. With permission.

PROTOCOL IMPLEMENTATION CONFORMANCE STATEMENT

The conformance statement is a document which should be provided by the manufacturer of the BACnet-compatible controller, PC, or another device identifying its individual features and its conformance to the BACnet standard. Owners should review the conformance statements of BACnet devices prior to purchasing such devices to make sure that the devices meet the job requirements.

NETWORK SECURITY

Network security in BACnet is optional. Data integrity and confidentiality are provided by peer and data origin authentication by using handshaking, and by operator authentication using operator passwords.

However, security is an important aspect of any network. Owners and engineers should understand the trade-off between network security and data distribution over the network. There are third-party network devices, in addition to individual DDC systems, that can provide sufficient network security for a real-time facilities network. Password alone is insufficient in most cases, and should be used in conjunction with other network security devices.

NOTES

BACnet protocol is probably the most talked about protocol of the 1990s in the BAS industry. Since its beginnings in 1987, it has been promoted on the pages of various magazines, trade publications, and books. After its release in 1995, the expectations have turned into reality in a form of a published, official ASHRAE Standard. Wide acceptance of the protocol was assumed, due to the composition of the committee and participation of most of the control vendors either on the committee or during the review process of the standard.

I have researched the "reality" of support and use of the protocol by interviewing some of the major DDC vendors. While all of them claim compliance with BACnet, none of them could show a "full-scale" installation of the protocol. This was a surprise after 4 years of existence of the protocol. When pressed from a "can do" stage of the discussion to "did you do it, where and how?" it was, in most instances, like talking about a distant person by somebody who heard about him from somebody who claims to know him.

Checking out their BACnet conformance statements, I found conformance classes ranging from Class 2 to Class 4. In reality, it means that if one wants to network three DDC systems, Classes ranging from 2 to 4, conformance Class 2 will be the common dominator — which supports only read and write applications services. The user will end up paying a premium for a protocol that does nothing more than an industry "de facto" protocol (i.e., Modbus) that can be purchased for a fraction of the cost. I also found more willingness of the DDC vendors to interface their systems to a SCADA type of front-end (i.e., Intellution, Wanderware, etc.), rather than to a front-end of their DDC competitor.

The BACnet protocol has all the features to become a true "standard" protocol. However, at present, building owners should be aware of the situation, and explore all the available options on the market. They should explore not only the availability of other protocols, but also the vendor's experience with installation of the proposed interface options, conformance of the proposed protocols or drivers with the owner's needs, support by the local branch office, and most important, the cost associated with the protocol and all related applications engineering.

13 Institutional Energy Metering: Challenges and Opportunities

Roberto Meinrath

Real-time energy metering opens the door to energy consumption analysis and management. The passive monthly readings of consumption traditionally associated with mechanical meters are now replaced by active real-time energy measurements, including flows, demand, electric phase balance, chilled water differential pressures and temperatures, condensate return temperatures, alarm reporting conditions, etc. Networked electronic meters open the door to integrated management of the production (one or more power plants), distribution (miles of underground utility systems), and consumption of energy (in dozens and even hundreds of buildings). Institutionally, one faces the difficulties inherent both in integrating traditionally separate isles of automation and in changing the institutional culture towards one of energy awareness and interest in managing and reducing building energy consumption.

In industrial settings, energy metering has tended to be part of the "floor" operation — in industrial processes, energy meters normally are quite visible to factory workers and their supervisors and managers. Even though a factory may be large, meters in industrial applications tend to be quite visible, within easy reach, and part of vast and integrated automated processes. In commercial and institutional applications, meters tend to be tucked away in building basements, out of sight and out of mind. While real-time metering and controls have for years been a standard feature in industrial processes, in commercial and institutional applications monthly readings of mechanical meters have been the norm. For most commercial and institutional customers, energy consumption tends to be an incidental cost and of secondary concern. In real estate properties, leases tend to include most utility costs and customers tend to be less than interested in daily consumption, trends, peak demands, etc.; energy costs are either part of the rental cost or tend to be a small percentage of the operating costs. For institutional consumers, such as a university or a hospital, department heads pay little or no attention to energy costs, as utility costs have tended to be managed centrally by a service department.

Real-time metering within a network application provides an opportunity for both service departments and customers to become energy managers. A service department, for instance, can now operate a power plant more efficiently by analyzing what is happening to the buildings — the power plant has become the heart of a large system, and the heart should respond to pulses received from the outlying buildings. No longer is a power plant sequence of operation defined exclusively by equipment efficiencies inherent in the plant itself: proper equipment sequences must also consider system-wide effects. Customers, on the other hand, can for the first time see the effects on energy consumption of variables such as weather, changes in operations (labs, etc.), day/nighttime and weekday/weekend operation, etc., all readily transparent to users of real-time energy metering systems. The challenge lies in making the management of energy consumption interesting and exciting — most probably synonymous with making it budgetarily attractive. However, one must first address the issue of reliability and credibility of real-time meters.

Real-time metering is both an engineering and an operational challenge. Engineering-wise, the challenge rises from defining the best meter locations to overcoming the difficulties inherent in communication networks, to integrating different metering applications into a common system, to integrating metering with other facility applications. In most institutional applications, meters are spread out over dozens and even hundreds of buildings that are not necessarily easily accessible and that pose problems ranging from identifying proper meter locations to designing a complex communications network. Once these engineering challenges are overcome, the engineers must then turn their attention to ensuring daily that "hidden" meters and their communication network are working properly. A commercially credible, viable, and reliable real-time energy metering system demands a repetitive and close-to-bureaucratic attention to daily details, with few if any engineering challenges — a significant operational challenge to the engineers and, thus, to the service department responsible for the metering system.

Once the operational challenge is overcome and an institutional metering system is reliable, the challenge shifts to the energy users. Institutional department heads, though always on the lookout for more institutional funding, are loathe to deal with the bureaucratic demands of the position. One must, thus, simplify any energy savings-related incentives. First, departments should not be affected by the volatility of fuel prices — a standard energy unit should be defined, with the central service department bearing all of the consequences of fuel price variances. Second, system engineers must spend some amount of time reviewing operating standards such as nighttime/weekend temperature setbacks, air flow controls, the PC-based controls available to the department, system alarms, etc., preferably with a departmental building manager, business manager, or equivalent, sparing the department head from such technical details. Third, system engineers must review with such departmental representatives the historical consumption patterns and trends upon which to base the energy incentives.

Institutional energy incentives could lead to relatively large savings. Since institutional culture changes slowly, energy-related savings should be worth the effort and should be spread out over a number of years. *The objective should be to permanently change energy consumption habits, not just to achieve one-time energy savings.* Since most of the energy savings would tend to be front-loaded into the first or second year of the program, an effective program should gradually reduce the sharing of such savings while still fostering the search for incremental savings. An approach, for example, might be first to establish reasonable energy consumption benchmarks, given prior years' consumption and weather patterns. Reductions from the benchmarks could then be shared on an 80/20 department/central basis in year 1, 70/30 in year 2, 60/40 in year 3, 50/50 in year 4. During the 4th year, one would revise the benchmarks for the following 4-year period. Of course, the slope of the energy savings sharing curve and the time line for benchmark revisions should be determined up-front and institutionally.

A healthy energy system (again, production, distribution, and consumption of energy) should behave not unlike a healthy body, with daily nurturing (analysis), exercise (engineering), a healthy diet (production), no sudden starts and stops, gradual ramp ups and downs, and a subconscious daily routine embedded in one's life. Unless such routines are embedded in the fabric of an organization, the routines will tend to atrophy. It is perhaps the single greatest challenge of any organization to state its objectives clearly so daily routines are properly identified and systematically implemented. Proper energy management not only generates savings, it also leads to reduced pollution and an improved building and global environment. Ultimately, energy management is not a choice, it is an obligation.

14 Case Study: Yale Maxnet

Viktor Boed

CONTENTS

ABSTRACT   One winter morning, just like every morning for 10 years, my daughter and I took our dog Max out for a walk around Yale's sport fields. (By the way, isn't it amusing that dog owners walk their dogs in the same pattern, in the same area, same streets, day after day, for years, without ever noticing it?) Well, as usual, the dog was doing his doggy things, while we had our father-to-daughter (or vice versa?) conversation about her school, her friends, my work, and other topics cut out for the leisurely morning walk with Max.

The conversation turned to my work. I was talking about different ideas for the most suitable real-time network for the University. After a while, I noticed her puzzled look and I knew I had lost her in the conversation. Upon asking her where I had lost her, she replied without hesitation that she had never heard a more messed-up naming convention, what with all those BAC*nets*, Ether*nets*, ARC*NETs*, and whatnot "*nets*" — how could anyone tell the difference? How could anyone understand what the Yale network will be, out of all those "*nets*?" We got into a long conversation on how the Yale network would be a combination of several types of networks. That cleared up the situation! After some more discussion she told me, "Dad, from now on, whenever you refer to the future Yale network, call it simply *Maxnet*, after our dog, and I will know what you mean." And that was the birth of Maxnet.

So, now you know why you can't find a "*maxnet*" in any professional or scientific publication (not even at Yale). Nevertheless, the name stuck to the Yale facilities real-time network. You can find it on the web (with Max's picture), showing the systems connected to the real-time network at Yale, in the name of the room with the network servers (Max room), and in conversations and correspondence referring to the "Yale real-time network."

Epitaph: Max passed away in 1998 at the ripe age of 16; we don't know if he went to the doggy or people's heaven, since we always suspected he thought of himself as human.

GENERAL

Yale Maxnet is a collection of systems networked together to provide real-time information for the operation, maintenance, engineering, and management of Yale's facilities. Since Yale has one power plant, one cogeneration plant, and a rather large number of buildings spread over a substantial area of downtown New Haven, the real-time communication network, Maxnet, is fairly large. Geographically, Yale occupies three areas, namely, the sciences, central, and medical school areas.

The various systems connected to Maxnet are the building automation systems (BAS), the power plant controls and automation systems (C&AS), and the Campus Building Utilities Metering Systems (CBUMS). More accurate description of the connected systems will follow.

Another characteristic that shaped Yale's network development is that the connected automation systems were implemented over several decades (for example, the BAS installation began in the mid '70s).

Network interfaces and the Maxnet itself were also developed over several years. The network development was influenced by the state of technology, the financial resources available for networking, and by our own engineering ability. The network is "home-grown," utilizing commercially available network components. Our goal was to provide interoperability and utilization of systems spanning several generations while preserving the investments in the existing automation systems and networks.

The biggest challenge for networking of existing systems is to engineer solutions that provide interfaces both to previously installed systems, thus assuring their continuous expansion, and to future network development, with minimum modification or obsolescence of existing systems and network components. We pretty much met that challenge.

BUILDING AUTOMATION SYSTEMS (BAS)

Building automation systems have a long tradition at Yale. BAS have been implemented continuously since the mid 1970s.

DELTA 2000

The first system implemented at Yale was a Honeywell Delta 2000 system for remote control and monitoring — a rather ambitious number, considering the system did not last into the 1980s. Until the era of direct digital control (DDC) systems, pneumatic single-loop controllers controlled HVAC systems. The Delta 2000 was a supervisory system, which was installed to monitor and control certain functions of the HVAC systems and/or pneumatic controllers (Figure 14.1).

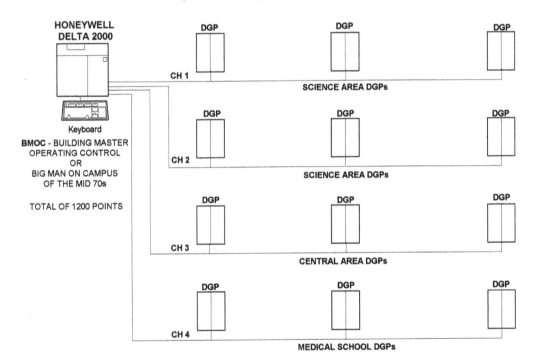

FIGURE14.1 Delta 2000 BAS.

The basic building blocks of the system were the data gathering panels or DGPs with individual point modules, multiplexing cards, and communication modules. There were some 1200 field points connected to the DGPs. The building DGPs communicated on 50- or 75-Ω coaxial cables to their front-end unit. Initially, Yale had four trunks in the central, science, and medical school areas of the campus.

Operators in the Building Master Operations and Control (BMOC or big man on campus — another ambitious name of the late '70s) office could address each field point individually from the Delta 2000 front-end. The Delta 2000 unit displayed real-time information (i.e., status or values), operating conditions (i.e., on line/off line), and could issue commands (start/stop), or re-set the set point of Honeywell pneumatic single loop controllers.

Delta 2000 Upgrade by EMS

Requirements for energy savings as well as advancements in energy management systems (EMS) technology set the stage for implementation of computerized EMS in the late 1970s and early '80s. Yale responded by upgrading the existing Delta 2000 system by a Steafa EMS-1 front-end. The EMS program ran on a PDP-11 minicomputer, with an interface to existing Delta 2000 DGPs developed by Steafa. The original Delta 2000 front-end was used as a back-up in case of failure of the PDP computer, the EMS program, or communications. The EMS system interfaced to the Delta 2000 was one of the first attempts of the industry to provide an interface to third-party systems (Figure 14.2). This was also an attempt to migrate older generation systems into newer generation front-ends, providing more sophisticated programs, operator interfaces, and energy management programs.

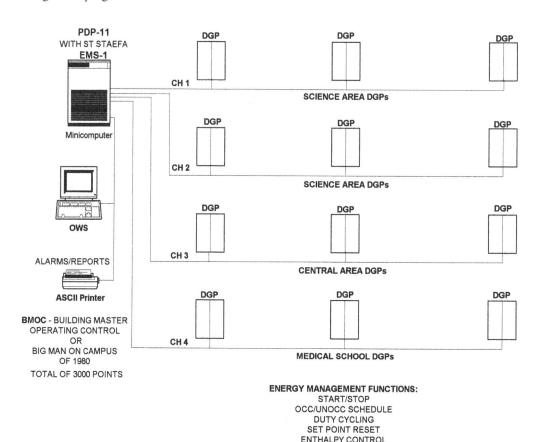

FIGURE 14.2   EMS interface to Delta 2000 DGPs.

The EMS system allowed for operator interfaces via several virtual terminals (VT) located in the BMOC office. The EMS program provided basic energy management functions, such as scheduled start/stop of air handling units (AHUs) and other connected equipment. Other programs, like occupied and unoccupied schedules, duty cycling, demand limiting, etc., were to modify the start/stop time of the HVAC system in an effort to save energy. Besides the start/stop functions, the system allowed for reset of set points of the connected Honeywell pneumatic single loop controllers. It also provided damper control (free cooling or economizer) based on enthalpy. The EMS system operator interface provided report generation and continuous alarm reporting.

Most features of the Steafa EMS system were utilized at Yale. As we learned more about BAS systems operation and energy management, some of the energy management features were reevaluated and consequently deleted from the system. The upgraded system grew into a fairly large BAS system over time.

The EMS/Delta system was Yale's flagship of automation for several years. As the requirements for controls and automation grew, so did the number of points connected to the system.

DELTA 2000 UPGRADE BY PEGASUS

In the mid '80s, the trend in building automation was moving towards implementation of distributed DDC systems. The University faced the dilemma of expanding the existing system, or replacing it with DDCs. Faithful to our traditional "there is a right way, a wrong way, and a Yale way," we selected the third option, and upgraded the EMS with a new PC front-end. The upgrade was done by a Pegasus system (from CENTAURUS, Inc.), which also provided some distributed processing capabilities (Figure 14.3).

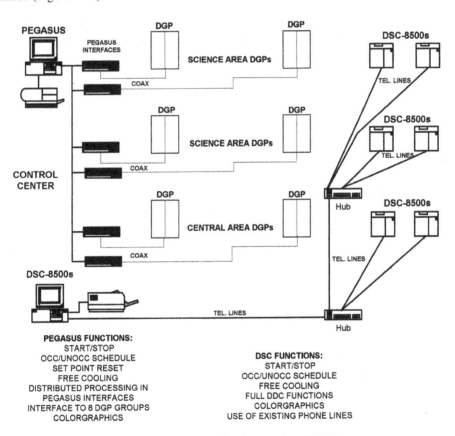

FIGURE 14.3   System upgrade by the Pegasus and implementation of DSCs.

Along with the upgrade, we decided for installation of new DDC systems in all new or retrofit installations. This assured preserving previous investments in automation (and our collective know-how in the existing systems), and provided retrofit and new installations with a state-of-the-art distributed DDC system for remote control and monitoring.

Upgrade of the EMS by a PC-based Pegasus system was also a prudent decision from a financial standpoint for several reasons:

1. It extended the life span of the installed BAS by almost a decade.
2. It saved investment dollars into BAS upgrade in buildings which were slotted for renovation over the next decade or so.
3. It avoided the high maintenance costs of the PDP minicomputer, which became increasingly expensive toward the end of its useful life.

DDC Systems

DSC 8500

Yale Facilities decided to open the BAS field to bidders, but retain control over the selection of the DDC systems and vendors. Five buildings were selected for design or retrofit of their mechanical systems. From a fairly large pool of DDC systems and vendors, the evaluation committee of Yale facilities ranked the preselected systems and vendors. (For vendor preselection, see also a book by Viktor Boed: *Controls and Automation for Facilities Managers — Efficient DDC Systems Implementation*, CRC Press, 1996.)

Yale was considering large-scale networking of real-time systems even before the BACnet committee was formed. One of the requirements was to preselect DDC vendors with open protocols so that other DDC vendors could write interface drivers or protocols to that system — a rather shocking requirement at the time of proprietary protocols!

Since the preferred selected vendor refused to open up their protocol, we moved to the next choice, which was Johnson Controls, Inc., DSC-8500 system. Just to illustrate how big a deal it was back then, here is a good story I wish to share with you. From the beginning of the preselection process we emphasized the requirement for open communication and interfaces to other systems. Our preferred vendor said they had the best system, the best support, the best network, we did not need interfaces to other systems. We insisted on networking. They stated they would convince us otherwise, and flew their beautiful, luxurious, fully equipped and stuffed private jet into New Haven. With the company brass and all the comfort of a private jet, they flew us into their home office. We got the VIP treatment, including limousines with uniformed drivers. When it came to the question of opening up their protocol, the head office confirmed their policy to support their own protocol, and had no plans for interfaces to third party vendors.

The story illustrates the "strong" mind set of the DDC vendors toward open communication in the mid 1980s, and the desire of many end users like us to have communication options.

Yale continued with installation of DSCs for several years. Our positive experience confirmed our decision for networking and DDC systems implementation. The installation also included the use of electronic/digital field gear, including electrical actuators for all new installations. We've stayed on this course ever since.

DDC Metasys

Introduction of JCI Metasys (and systems like that) prompted us to evaluate the DDC systems on the market yet once again. This resulted in new DDC system preselection for mechanical retrofits of seven residential colleges. From four evaluated vendors, we selected another Johnson Control DDC system — Metasys. Based on our previous positive experience with the local JCI branch office and the JCI system, Yale entered into a partnership agreement with Johnson Controls. This included the lowest pricing from JCI, open-book pricing, applications engineering assistance throughout all design phases, and 1-year system performance warranty on installed systems, including testing out the operating logic through the four seasons. The sole-source agreement allowed us to develop a large and sophisticated system, and train our engineers, operators, and technicians to a high level of proficiency. We also developed design standards and highly sophisticated control routines, which were implemented uniformly on all installed systems. Over time, we migrated and interfaced all existing BAS systems into the Metasys (Figure 14.4).

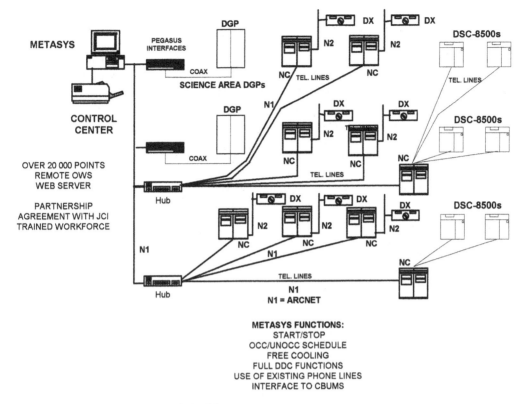

FIGURE 14.4   BAS system upgrade by Metasys.

JCI was able to implement a large scale ARCNET over our existing dedicated phone lines, thus saving expenses for dedicated communication trunks. The network grew over time into a large system with over 50 nodes — 50 network control units (NCs) — on the Metasys network. These building NCs interface to hundreds of unitary controllers in the buildings, controlling some 20,000 connected field points.

Over this past decade, due to its performance and reliability, the Yale BAS system became "just another utility." The system is reliable; it requires minimum maintenance (considering the high number of installed points), minimum engineering, and it controls and monitors just about every major building on campus. Building occupants as well as O&M personnel became accustomed to rely on it — this is probably the highest mark a system can earn over time.

Metasys Upgrade to Ethernet and Its Web Server

In an effort to optimize services and communications among facilities management and the real-time systems, the network underwent rapid changes. Requirements shaping the upgrade were

- **Technological,** such as advancement of so-called web servers, which allow users, within or outside the University, access to BAS data via the Internet.
- **Organizational,** which is the possible move of Yale Control Center operation outside the University to an external location. The CC in a new location would be interconnected with the University via fiber optic communications owned and operated by the local phone company.
- **Structural,** such as the decision of facilities management to outsource the operation and maintenance of selected buildings to outside maintenance organizations.

The following system and network modifications were implemented (Figure 14.5):

1. Conversion of the existing Metasys ARCNET communication to Ethernet.
2. Designating a University Metasys *archive* station with archive data from buildings maintained by the University as well as from buildings operated and maintained by outside organizations, connected to both ARCNET and Ethernet communication networks.
3. Designating a Metasys web server for the University, which would provide information to inside and outside users over the Internet.
4. Setting up a Metasys OWS with a web server for the third-party O&M organization in one of the buildings operated and maintained by that organization.
5. Providing an OWS with a third-party software (pcANYWHERE), in buildings under third-party maintenance contract, to mimic the Metasys screens.
6. Providing the new buildings connected to the Ethernet with Ethernet network control units.
7. Setting up the Ethernet Metasys OWS in the new Control Center.

This arrangement provides several advantages for the University:

- Preservation of investment into existing systems.
- Preservation of integrity of the existing Metasys system on the ARCNET, which is very reliable and well maintained.
- Compartmentalization of systems and responsibilities assigned to outside O&M organization (separation from the in-house systems and organizations).
- Ability to maintain a University-wide (Metasys) archive database, regardless of who maintains the buildings.
- Provision of full access to Metasys via Metasys OWS for the Control Center and controls mechanics for online monitoring and troubleshooting.
- Distribution of information to other users within the University, such as to zone managers, engineers, and facilities managers located all over the campus, via the Metasys web server. The implementation of web servers saves money for additional Metasys software licensees, while distributing information to the interested parties within and outside the University.
- Access to the systems managed by outside O&M organization from their business locations — outside of the University. By having a dedicated Metasys web server for each organization, the access is limited to the systems managed by that O&M organization.

OPENING UP THE UNIVERSITY TO MULTIPLE BAS SYSTEMS

Nothing lasts forever — not even a single-vendor BAS environment at Yale. Construction managers managing capital projects tended to blame their project overruns on a single vendor environment at Yale (not exactly a valid claim, considering the BAS costs a fraction of the overall project cost, usually less than the contingency for the entire job). They repeatedly requested that Yale open up capital projects for competitive BAS bidding to obtain the lowest first-cost for BAS systems.

Yale has decided for competitive BAS bidding for all capital projects and major renovations. Fortunately, by this time, all BAS systems were migrated onto Metasys, and the power plant and metering systems, including a Metasys DDE interface, have been migrated into Maxnet. Yale was ready for integration of its BAS and for multi DDC vendor participation for its capital projects.

Networking protocols offered by most BAS vendors seemed to be, at this time, LonTalk® at the building controller level, and BACnet at the higher level. Even though most DDC vendors were promoting these options, their branch offices have had very little or no experience with networking

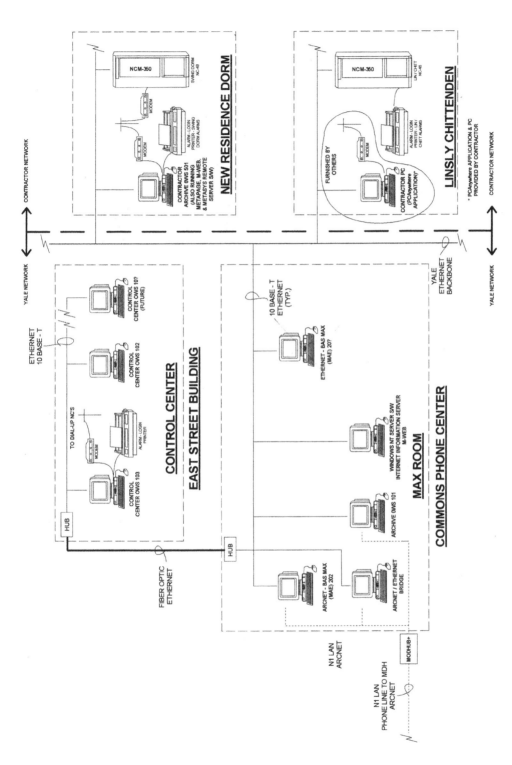

FIGURE 14.5 Metasys upgrade by Ethernet and Web servers.

with other DDC systems. Another major obstacle was the mindset of the vendors, conditioned for many years by competitive bidding.

I wish to bring up a few items, since the question of single or multiple BAS vendors is in the minds of most facilities engineers and administrators.

- The first costs of BAS systems are a fraction of their life cycle cost (similar to mechanical systems). Therefore, life cycle cost, not first cost, should the determining factor in making a decision.
- Besides the obvious financial aspects of the decision-making process, one should also consider operating and maintenance aspects, such as the proficiency of the O&M personnel in maintaining one vs. multiple DDC systems. The financial gain from bidding can be easily spent on service contracts and spare parts from multiple BAS vendors.
- Open bidding requires more detailed control specifications, sequence of operations, controls drawings, etc. from the designers of mechanical systems — an expertise most mechanical engineering firms lack (to various degrees).
- Is the money saved on the first cost for a DDC system spent on the first cost of additional networking and first cost for other building systems? At Yale, for example, with the start of a multiple-vendor environment, the metering system and its communications had to be excluded from the DDC bids, since it is not known which system is going to be selected for any particular building. Until recently, metering systems were included in the Metasys DDC system to which we had an interface. In fact, the points for metering were connected to the DDC field panels, which already had network connection to the metering system's server, thus saving money for field installation.
- Besides BAS considerations, one also has to look at networking issues. How is the multivendor system going to work on existing communication networks? Does one need to set up another network, bridges, routers, gateways, or other network configuration? Here too, one has to look, besides the first cost, to network integrity and to possible additional costs associated with network operation and maintenance. Will such a network require additional maintenance and/or management? Will the network expansion require additional training, maintenance, spare parts, management, and other organizational changes and additional expenses? Will the proficiency of the network O&M personnel managing and maintaining multiple networks be the same?

These are real questions and real expenses. The savings from first cost can be easily wiped out (and exceeded) during a short period of time of the systems operation. Also, the reputation of system administrators could be easily tarnished, if a new system or communication problems would cause building problems. Customers and building occupants do not like experiments that inconvenience them, even if there is an up-front first cost savings.

Yale has had good experience with preselecting BAS vendors in the past. We decided to proceed once again with a BAS vendor prequalification process. From the pool of five vendors, we have prequalified two additional vendors: Siebe Environmental Controls and the Landis Division of Siemens Building Technologies, Inc., in addition to Johnson Controls, Inc.

The new multivendor environment has to meet the following goals:

1. Preservation of integrity of the existing system.
2. Communication on the Yale Ethernet fiber optic backbone at the campus level.
3. LonMark® communication within the buildings.
4. Compartmentalization of systems, by each system (a) having its dedicated OWS (for troubleshooting and maintenance) and (b) having vendor-specific buildings (no two BAS vendor systems in the same building).

5. The BAS systems have to have interfaces to Intellution Fix 32 (our Maxnet) and/or JCI's Metasys (our DDC).
6. All BAS systems must have dedicated web servers.
7. The selected systems and their communication options must be commercially available and must have been in operation for at least 1 year at a similar site. If the proposed system does not meet the 1-year operating experience requirement, the vendor must test the system and its communication at Yale for an extended time (i.e., 1 month).

INTEGRATED BAS

To move ahead with BAS system integration, we had to make the following decisions (Figure 14.6):

- We have decided to utilize the Metasys front-end OWS as an operator interface. JCI has introduced a new OWS upgrade with split screens and multiple CRTs connected to the same OWS-PC. This allowed us to retain the PMIs familiar to our operators. Each vendor specific system can be viewed on a dedicated screen if the operators desire to do so, or the CRTs can be utilized in any combination, including split screens on each of them. This provides a universal set-up with nearly unlimited options for the operators. The same PMIs (color graphics, etc.) are also used for the web server. This saves programming cost, since the graphics do not have to be developed twice.
- Building level communication, provided by each vendor, is a serial communication with LonTalk protocol. This allows interchangeability of network components and universal interface to third party controllers.
- Campus level communication among BAS systems is via the Ethernet campus network. Each vendor provides its own OWS-PC. The PC is used for the control mechanics to have full access to vendor-specific controllers, and in case of failure of the campus network, it provides access to each vendor-specific network. Each vendor-specific OWS-PC has a BACnet interface to share information with each other. In our situation, BACnet provides information from vendor-specific systems to the "front end Metasys OWS-PC."
- We need to have only one interface to transfer information from the BAS systems to the Maxnet. This allows full utilization of the existing networks and interfaces without the need to redevelop them.

POWER PLANT CONTROL AND AUTOMATION SYSTEMS

Sterling Power Plant

Sterling Power Plant (SPP) provides steam and chill water to the Yale School of Medicine (YSM) and Yale New Haven Hospital (YNHH). YSM is a school of Yale University with its own administration, operation, and maintenance management. YNHH is an affiliated, independent healthcare corporation with its own management. SPP is operated and managed centrally by the Yale Utilities Department.

The Sterling Power Plant has five boilers on dual fuel (oil and gas), producing a total of 220,000 lb of steam. The high pressure steam is distributed to the buildings in underground steam tunnels, then reduced to a medium and/or low pressure, according to the building needs (Figure 14.7).

There are six chillers (steam or electrically driven) with a total capacity of 15,000 tons. The chill water is distributed in four interconnected loops to the buildings. In addition, the plant has a cool pool with 2.7 million gallons of water capacity, used primarily during summer peak hours. The cool pool can supply 750 to 1000 tons of chill water for about 20 hours. Another energy savings option is the use of a flat plate exchanger which supplies free cooling (500 tons) during the winter months.

The chill water system is an all-primary system with pumps in the SPP (Figure 14.8). From SPP, the chill water is distributed to individual buildings in the central and science areas. The

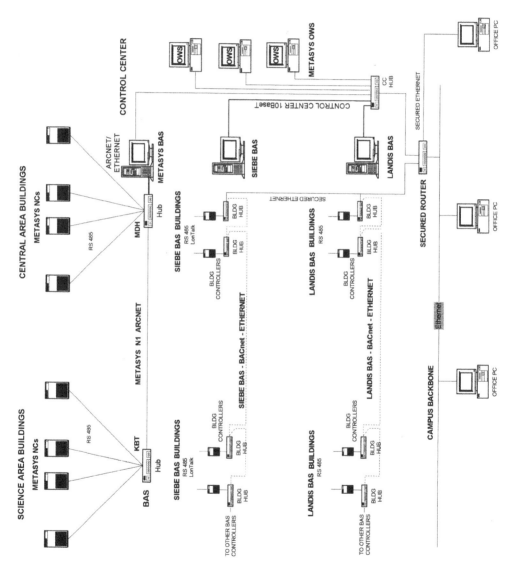

FIGURE 14.6 BAS integration.

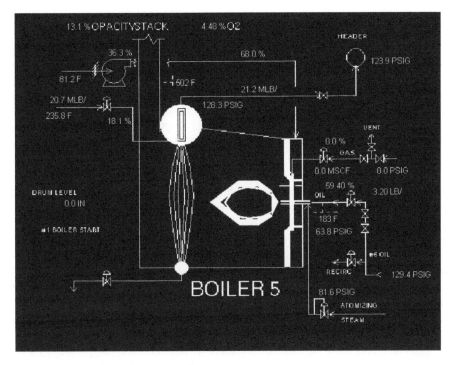

FIGURE 14.7 Example of a boiler number 5 screen.

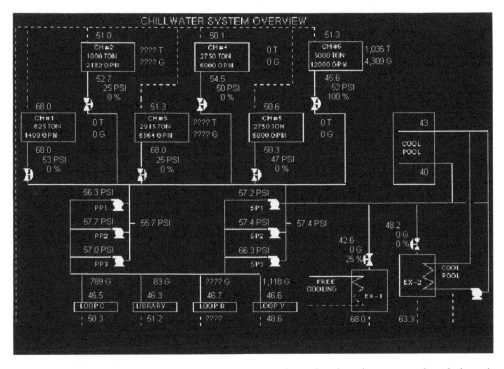

FIGURE 14.8 Example of a chill water overview screen (note that the printout was taken during winter with minimum chill water load).

all-primary chill water distribution system has no bridges and secondary pumps in the plant or in individual buildings.

The plant is controlled by a Yokogawa Digital Control System (DCS) called Micro Excel (MXL). The system consists of field control units (FCUs) connected via a redundant bus to operating work stations located throughout the plant and in the control room (Figure 14.9). Non-Yokogawa equipment is integrated into the FCUs via serial communications and Modbus drivers. The FCUs are set up with dual processing, dual communication modules, and battery backup, which provide very reliable operation and seamless switchover in case of a CPU or communications failures. All equipment (Yokagawa and non-Yokagawa) in the plant is mapped over the network and therefore visible from any operating station. The Micro Excel provides graphical display screens, trend graphs, detailed operating screens, and other person machine interfaces (PMI). For their convenience, the operators developed a substantial number of graphical screens. Besides control functions, the system also provides basic optimization functions, alarm reporting, and report generation.

The Micro Excel system is interfaced to a dedicated Maxnet server via serial communication and interface driver. All points (tags) from the system are mapped over to the server, and all graphical screens are available online on the Yale network and/or Internet.

The server also provides load forecasting for the boiler and chiller plants. The forecast is based on daily weather (temperature and humidity) forecasts. The system calculates the forecasted load by "looking back" at the history of a similar day and equipment (chillers and boilers) used. From the forecasted weather data and the historical data on boiler and chiller loading, the system then forecasts the steam and chill water load for that day, in 4-hour intervals. Based on the forecasted loads, the plant operators make the needed equipment (boilers, chillers, pumps) ready for operation. Even though this method is less accurate than simulation methods using linear equations, it is also much less costly and quite sufficient for equipment loading.

The data from the plant are distributed over the network to plant engineers and power plant management and to engineers of YSM and YNHH residing in remote locations.

Central Power Plant

The Central Power Plant (CPP) was modernized and converted to a cogeneration plant during 1996 to 1998. CPP produces steam, chill water, and electricity for the central and science areas of the campus. The steam is also used for cogeneration.

The steam plant has a stand-alone boiler and three heat-recovery boilers fired by oil and natural gas, producing 200,000 lb of steam at 225 lb pressure. The plant has three 6-MW gas turbines that generate electricity and steam as a by-product. The high pressure steam is distributed to the science area of the campus via direct buried interconnect steam distribution piping. For usage in buildings, the steam pressure is reduced in pressure-reducing stations to 60 and 10 lb. The steam distribution system in the central area is medium and low pressure (125 lb, 10 lb). The low pressure steam is used mostly in heating and HVAC systems, the medium pressure steam in the kitchens, laboratories, and other (Figure 14.10).

The condensate from the building is brought back to the plant, treated, and returned to the boilers along with the make-up water. Steam meters measure steam production. Building consumption is measured by ultrasonic condensate flow meters.

The chill water is produced in steam-driven chillers. There are 4 chillers in CPP, producing a total of 9000 tons of chill water. The chill water system is an all-primary system with pumps in the CPP. From CPP, the chill water is distributed to individual buildings in the central and science areas. The all-primary chill water distribution system has no bridges and secondary pumps in the plant or in individual buildings (Figure 14.11).

To provide pressure limitation for buildings closest to the plant, the chill water supply pressure is reduced by differential pressure stations (similar to the ones used in the steam distribution systems) to pressures that can be tolerated by the control valves of the building HVAC systems. For example,

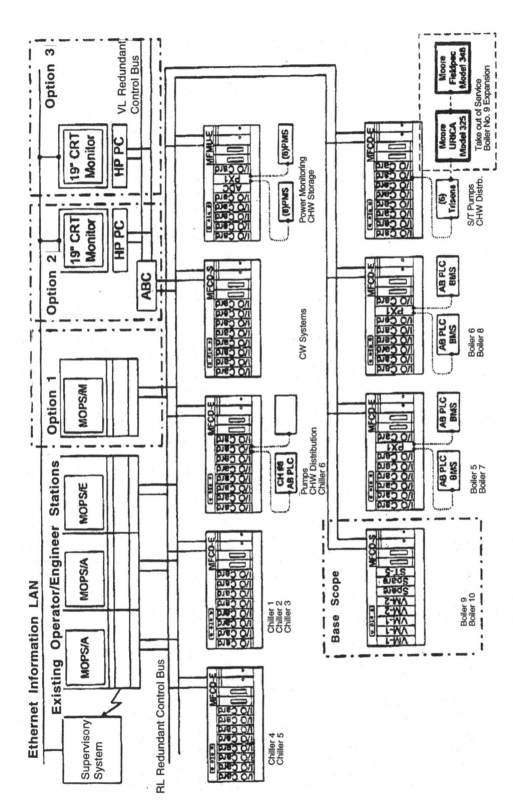

FIGURE 14.9 Control network at Sterling Power Plant.

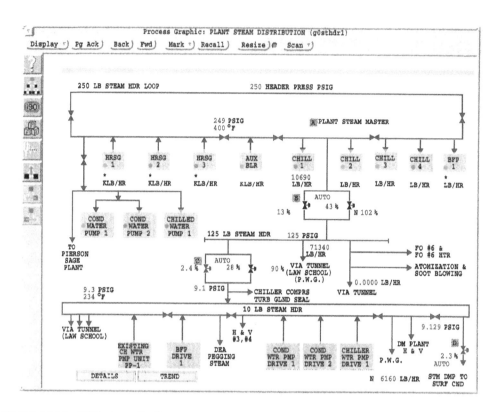

FIGURE 14.10   DCS steam distribution screen.

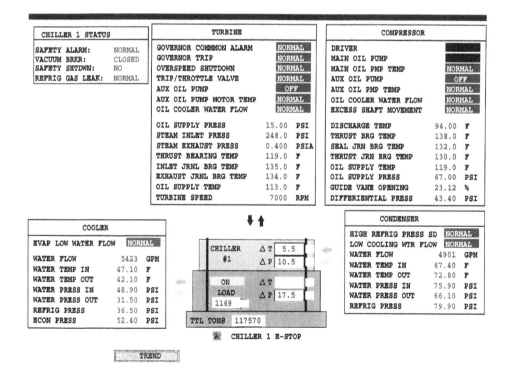

FIGURE 14.11   Chiller 1 control screen.

fan coil unit (FCU) control valves are sized for fairly low entry pressure (i.e., 30 lb) and low differential pressure (i.e., 10 lb). In buildings closest to the plant (high chill water supply and high chill water differential system pressure), the FCU control valves become uncontrollable. The pressure overcomes the shaft pressure (originated from the spring and electrical actuator) and lifts the valve, providing uncontrollable flow of chill water and consequent cooling by the FCUs. High pressure conditions occur with limited cooling load conditions in the buildings mostly during the winter. Uncontrolled comfort cooling may be tolerable during the summer, but it is a total waste of energy during the winter.

Another method to meet the building pressure requirements in an all-primary system is the use of flat-plate heat exchangers. They are used either to reduce pressure, or to separate the building from the system pressure in the case of the high-rise buildings (which require higher pressure than the distribution system can provide without unwarranted increase of the supply pressure for all buildings connected to the distribution system).

The pumping pressure is controlled by VFDs in the plant to meet the required pressure distribution for the connected buildings. Ultrasonic chill water meters meter each building connected to the distribution system. Supply and return temperatures and pressures of each chill water intake are monitored by the Central Building Utilities Metering System (CBUMS).

Cogeneration is provided by three 6-MW gas turbines. Since the high voltage normal electrical distribution system as well as the utilities companies system is at 13.8 kV, the generated voltage at 4160 V is increased to 13.8 kV. The total generating capacity of the cogeneration plant is 18 MW. Alternate and peaking power is also generated in the CPP by three 1.5-MW diesel generators producing 4.5 MW of peaking and back-up power.

The generators as well as the utility feeders are connected to a double-ended main switch gear (Figures 14.12A and 14.12B). The main distribution system cables connected to this switch gear feed normal and emergency power to electrical substations in the central and science areas.

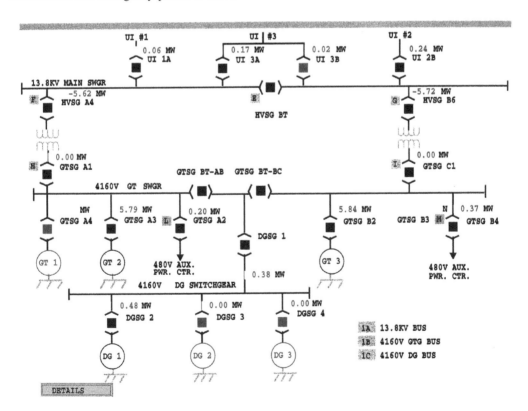

FIGURE 14.12A   CPP electrical/generation screen.

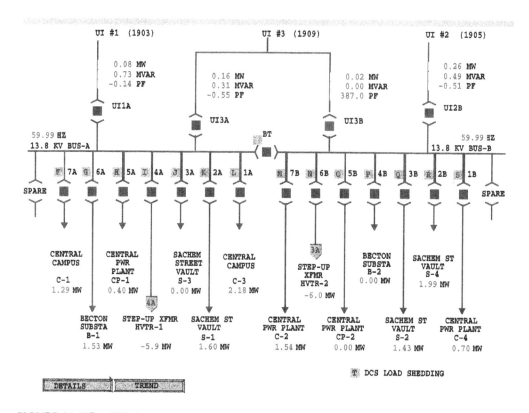

FIGURE 14.12B   CPP electrical switch gear.

The electrical generation section of the cogeneration plant is controlled by its dedicated GE control system (Figure 14.13). The overall plant control is by Bailey Distributed Control System (DCS).

CENTRAL BUILDING UTILITIES METERING SYSTEM (CBUMS)

"You can't manage what you can't measure," translated into energy-conservation language, means that you can't manage energy unless you measure it. In the past 20 years, there were many energy conservation programs based on other than direct measurement of energy. These methods have ranged from calculating energy usage based on square footage of radiation, to calculating savings from comparing energy ratings (i.e., kilowatts) of old equipment to retrofitted or new equipment. Such calculations were sufficient to obtain energy credits or so-called "energy incentives" from the government or utility companies. Nevertheless, the fact remains that unless one measures energy, one does not have a full understanding of energy consumption on the building or departmental levels. Furthermore, without direct measurement of energy usage it is difficult to define energy saving opportunities. The attempts to circumvent direct measurement were and still are (in the majority of cases) due to the high cost of meters and their installation.

INDUSTRIAL FLOW MEASUREMENT AND METERING ELECTRICAL CONSUMPTION FOR BILLING

Meters for flow measurement used for energy metering are often the same as the ones used for measuring liquid in industrial processes (fps, GPM). Electrical consumption (kWh) in most applications is measured for billing purposes.

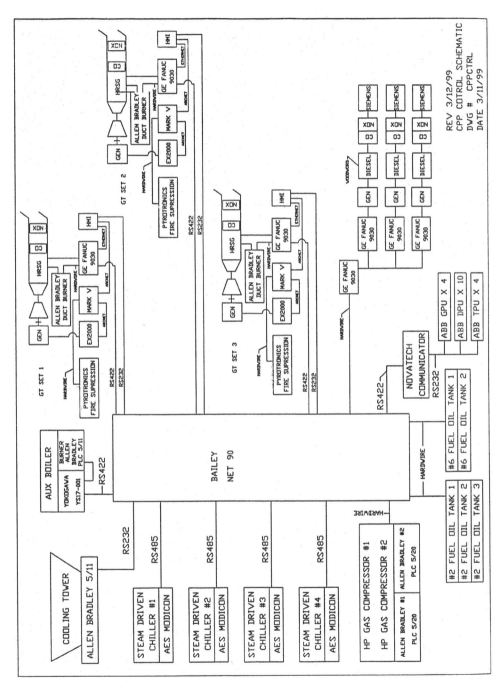

FIGURE 14.13 CPP control schematic.

Measuring energy is different from measuring flow. Let's look at industrial flow measurement. In an industrial process, flow is within a defined range determined by the process itself. During production, the flow will remain within its defined range. When the production process stops, there is no flow to measure. Therefore, the turn-down ratio of a meter is small, matching the flow fluctuation of production.

Another important factor is that the production piping is sized for particular flow conditions and for optimum flow velocity. This assures near-optimum flow velocities throughout most operating conditions. The small turn-down ratio and optimum flow velocities in the pipes are important for sizing, selecting, and locating flow sensors.

Measuring electricity is, in most cases, associated with measuring kilowatt-hours (kWh) for billing. Many metering systems are based on taking the pulse output from utility kWh meters, and compiling the totaled kWh into history files and spreadsheets. Such a method is sufficient for most metering applications for billing.

Instrumentation for flow and electrical metering is always supplied (and installed) by their respective vendors. Most systems provide local readouts with an option to communicate to their own front-end PC, using proprietary communication protocols (Figure 14.14).

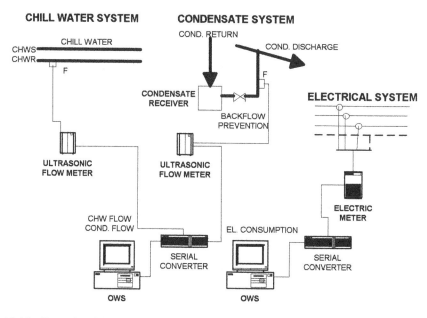

FIGURE 14.14   Example of flow and kWh measurement.

ENERGY METERING

Flow metering or **energy metering** for utility applications is different in that energy from a central point (power plants or utilities intake points) is distributed via the distribution system throughout the plant or campus. The flow in a chill water distribution system varies from the maximum, during high peak load conditions, to the minimum flow during off-peak conditions. Flow velocities in utility pipes, even at peak load conditions, are much lower than in industrial systems. Utility distribution systems are designed for maximum known load conditions (at the time of design). The distribution system piping is then further oversized for the estimated future loads. In addition, design engineers consider a safety factor and select larger pipe sizes for the designed distribution system.

This results in low velocities in the distribution piping. For example, the velocity in the pipes of a chill water distribution system is at its highest during summer peak loads, which are of relatively short duration. This means that the velocities in the pipes are way below their design conditions most of the year. The flow is further reduced during night setbacks, shoulders, and off-peak operations. To match the load profile of such a chill water distribution system, the selected meter has to have wide operating range, good turn-down ratio, good repeatability, and accuracy.

For energy calculations, we have to measure other variables, such as chill water supply and return temperatures. The accuracy of energy calculation depends on many factors besides the accuracy of the flow meter and temperature sensors used. The end-to-end accuracy is reduced by factors such as accuracy of the flow computer doing the calculation, scan time of the field sensors, and scan time of the communications driver running between the flow computer collecting the information and the associated network server. Another factor reducing the accuracy of energy calculations is, for example, using present values (PV) from field points connected to different field panels (or flow computers) scanned at different scan rates in the same equation.

Besides energy calculations, metering systems can also provide other useful information for optimization and troubleshooting for facilities engineers. For example, pressure measurement of buildings connected to the chill water distribution system can be used to monitor and/or troubleshoot the hydromics of the distribution system and the connected buildings. The chill water plant can also use such pressure values to optimize pumping horsepower to meet the changing load conditions in the course of a day.

Electrical metering for utilities applications is called "**power metering**," which represents measurement of voltages and amperes of the electrical system, and calculation of the required variables, such as kilowatts, kilowatt-hours, kilo-vars, power factors, harmonics, and other electrical data. Power measurement systems provide much more data than just kilowatt-hours for billing. They also provide online set-up parameters, and data for management and troubleshooting of the monitored electrical distribution system.

Facilities energy metering systems should be designed to provide data to facilities managers and engineers over a "common" communication network (Figure 14.15).

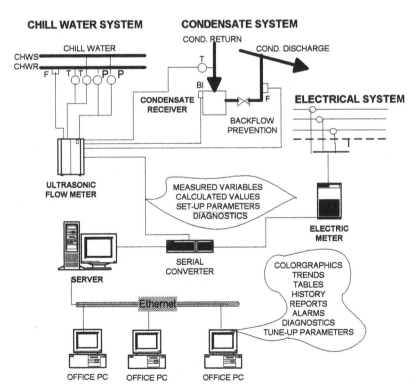

FIGURE 14.15 Example of facilities energy metering.

GOALS FOR YALE CBUMS

The Yale CBUMS was designed to achieve maximum return on investment by providing data obtained from the metering system to the end users in several ways.

- **Instantaneous values** from the field instruments and/or calculations are displayed to the users in the form of color graphics, tables, and trend graphs. They represent the core information on systems operation to engineers monitoring the system. They provide instantaneous readings of system parameters to the engineers, BAS operators, and power plant operators. Selected data are electronically transferred from one system to another. For example, chill water pressure values from the metering system (CBUMS) is used to make set-point adjustments to the variable frequency drive (VFD) of the chill water pump in the power plant.
- **Alarm reporting** is an important function of every system. Alarm reporting from CBUMS is designed to advise operators on system problems — for example, low voltage or high current of the metered electrical system; low differential pressure at the building intake or high temperature of the chill water entering the building. The metering system also provides more sophisticated alarm reporting, such as phase overload and phase unbalance of the electrical system. Such alarms can prevent electrical failures or brown-outs, by providing early warning for the maintenance department on a pending problem. Another example is a high condensate temperature alarm advising the maintenance department of a failure of steam traps.
- **Performance analysis** of the following systems:
  - Monitored distribution system, such as analyzing the pressure and flow relationships of the chill water distribution system at each intake point; power factor of the electrical system, etc.
  - Building performance; for example, analysis of flow, pressure, and temperature conditions of a building can lead to improved operation of the connected HVAC systems; a negative reading of a condensate flow points to a faulty check valve.
  - Metering data; for example, signal deterioration of an ultrasonic chill water flow meter could mean loose transducer connection or aeration in the pipe.
  - Connected network and computers by running diagnostic programs installed, for example, on the network servers.
- **Report generation** of the CBUMS is divided into several categories:
  - Meter reports, providing hourly (or per demand interval) readings associated with individual feeders.
  - Building consumption reports, designed as monthly reports used for energy (or cost) allocation and billing. The building data can be taken from an individual meter (if the building is metered by one meter) or can be a result of calculations, if there is more than one feeder or meter associated with that particular building. For example, the data can be a summation if the building is fed by more than one feeder or subtraction if the flow is measured before and after the building.
  - Special reports used, for example, to measure energy usage of buildings, to trouble-shoot systems performance, to compare consumption of buildings of the same kind (i.e., compare consumption of chemistry laboratories), current consumption to previous consumption of the same period (i.e., current month to the same month last year), etc.
- **Electronic transfer of data** to users, central data warehouse, energy accounting systems, or other systems.
- **Providing data for the users via a public web server.** The data can be compiled for each building in a tabular or graphical format.

YALE CBUMS

Implementation of CBUMS began with installation of electrical and chill water flow meters in the buildings. The meters installed in this phase were in the central and science areas of the campus.

ELECTRICAL METERING

Survey and Design

Prior to installation of the power metering system, the existing electrical distribution system and associated switch-gear were surveyed. Based on survey results, the following actions were taken:

- One-line electrical distribution system diagrams were updated and used with minor modifications as color-graphic screens
- Power meters were located on individual switch-gear so as to meter individual feeders and/or buildings on every voltage level (Note: (1) not every feeder had to be metered, (2) some unmetered feeders and/or building consumption were calculated)
- Design parameters were double checked for the electrical systems, their configuration (Delta, WYE), voltages, phases, and currents (important for setting up meter parameters) (Figure 14.16, see next page)
- The metering systems engineering, sizing of current transformers (CTs), voltage transformers (PTs), fuses, and shorting blocks, as well as design of their installation to enclosures or into switch-gear panels (Figure 14.17, see page 217) (some meters were installed into enclosures, others on the switch-gear panels)
- Engineering of the communication circuits and assignment of meters to each circuit (the number of meters and the length of each circuit is the most important parameter of serial communication)
- Setting up converters from RS-485 signals to RS-232 prior to connecting individual circuits to serial ports of the server

Installation and Startup

Based on the in-house engineering, the meters were ordered, electrical installation was completed, and the associated software and communication drivers were programmed.

Upon completion of the electrical installation, software development, and on site testing of the meters and communication, the system commissioning was done in steps:

- The electrical installation was checked out locally (from the PCs and CTs to the meters)
- Individual meters were set up and programmed either locally (older meters), or via the network (newer meters) resulting in reading the data locally on the meter's LCD display
- Communication circuits were checked out from the telephone hubs and from the server
- Upon reading the meter set-up parameters and "credible" electrical data on each meter locally, the meters connected to the circuits were scanned and the database in the server was updated with real-time data

At the time of the first electrical metering installation, our selected vendor, Power Measurement, Ltd. (PML), was supporting proprietary as well as Modbus drivers for serial communication. Since the Modbus driver was a standard for serial data communication for Yale installations, the PML meters were purchased with the Modbus communication option.

As for the physical media, either dedicated phone lines or direct wiring was used. In using the phone line option, the meters are connected to telephone break-boxes and to their respective telephone circuits. Assignment of meters to phone circuits required special attention so as not to

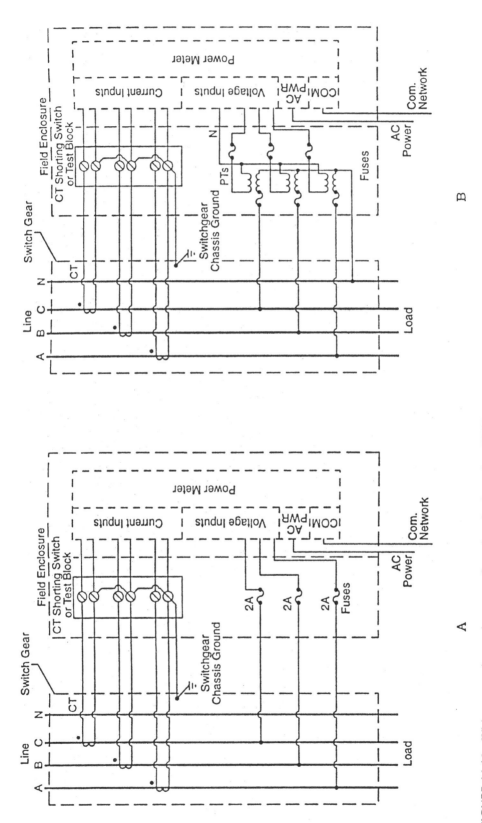

FIGURE 14.16 Wiring connections and meter configurations for three-phase WYE systems: (A) three-phase, four-wire with direct voltage; (B) three-phase, four-wire with three PTs and three CTs. (From Boed, V. *Controls and Automation: Applications Engineering*, CRC Press, Boca Raton, FL, 1998. With permission.)

3 Phase Connections

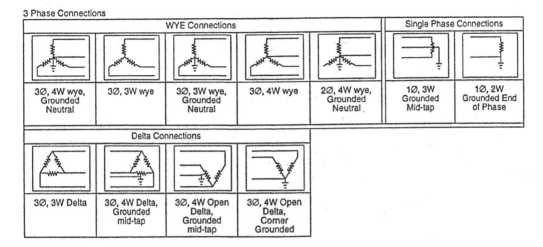

FIGURE 14.17 The most common electrical power system configurations. (From Boed, V. *Controls and Automation: Applications Engineering*, CRC Press, Boca Raton, FL, 1998. With permission.)

exceed the maximum allowable cable length for the connected number of meters for each RS-485 circuit. At the server location, the telephone lines (or the proprietary wiring) are connected to a bank of serial converters and, via RS-232 communication lines, to the communication ports of the server. Each circuit has a designated driver in the server polling the connected meters at predefined time intervals (Figure 14.18, see next page).

Data Presentation

Data presentation is by means of **color-graphic screens**, accessible either from the menu or from the electrical overview screens for each University area (Figure 14.19, see page 219).

Each screen provides schematic representation of the related electrical switchgear, with the meters shown on the metered feeders (Figure 14.20, see page 219). Limited, essential meter readings appear in "dialog boxes" (named "dynamos" in the Fix 32 system). Clicking on the dialog box brings up a more detailed data field (on the right-hand side of the screen) divided into four sections: Voltages, Amps, Power, and Demand. The data in these fields are read at a scan rate from the registers of individual meters. Engineers and facilities managers evaluating building and system performances use the screens daily. The screens have become important for evaluation of building operation and maintenance, energy management, and for providing real time engineering information for new design.

The screens are set up in a "tree" structure, using "soft buttons" to move from screen to screen. Part of each screen is dedicated to "navigation buttons," which are used to move from one screen to another, from one system to another, or from one area of the University to another.

Each area screen is set up with relevant telephone numbers to engineering and O&M departments, and with color photographs representing a view of that area. By means of soft buttons, one can move to other areas of the University or select the desired system within the displayed area.

The lower part of each screen also contains an "alarm window," which displays all recent alarms. The user can get to alarm details by accessing the alarm screens.

Another representation of real-time data is by means of **trend graphs** (Figure 14.21, see page 220). Trend graphs provide continuous representation of the data assigned to each of the eight "color pens" for each trend graph. The trend graphs provide not only real-time readings, but also history data in one continuous graph. Trend graphs are used for analysis of systems performance, of distribution system problems, as well as for future design.

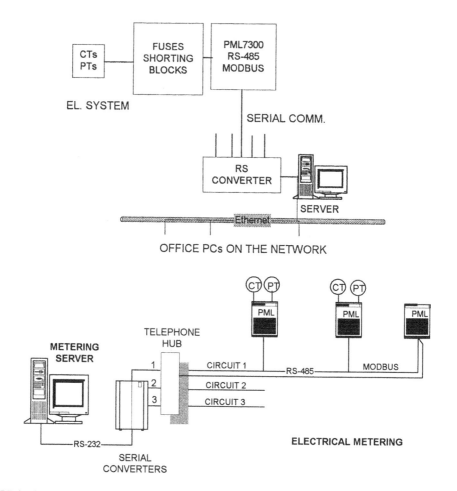

FIGURE 14.18   Connection of the electrical meters to the network: (top) principal diagram, (bottom) circuit diagram.

Reports Generation

Standard Reports

Report generation is one of the important features of the system. PML meters provide calculations of kWd (demand) and kWh (consumption). CBUMS servers scan the connected meters and fill up the database with 15-min kWd readings. Each server then calculates the hourly consumption for the connected meters. Demand and hourly consumption data are compiled into so-called "daily reports." The 15-min readings represent the core of the daily reports. Meter totals from daily reports are used to compile monthly reports for individual meters.

Meter reading and statistical data validation are done in several steps:

First, the computer interpolates the missing data of each meter from the "good" readings. Missing data may be one of two kinds:
  • No data are available from the meter due to meter or communications failure, or
  • There is no consumption at the time of the reading (for example, the feeder is not used)
Second, engineers responsible for engineering oversight of their areas and systems review the meter readings and provide validation of the reported data. For example, there may be "bad" data reported, due to mismatch of meter parameters with the parameters of the electrical system (Figure 14.22, see page 220).

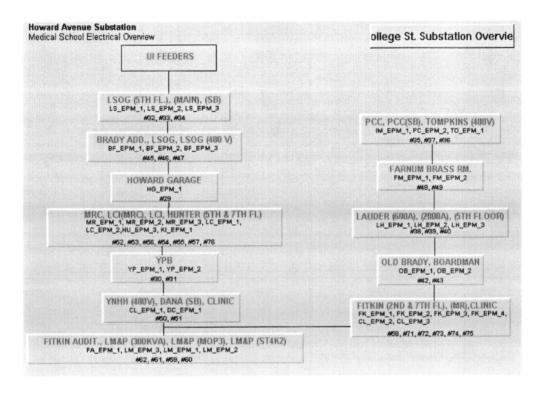

FIGURE 14.19 Example of an electrical overview screen.

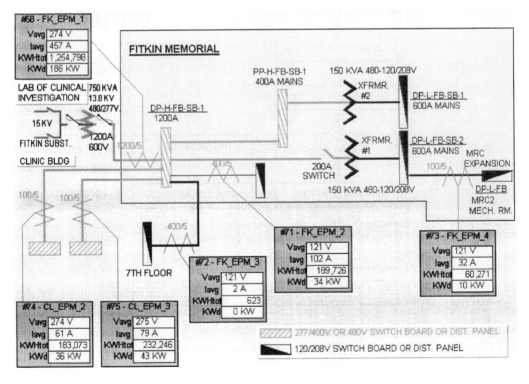

FIGURE 14.20 Example of an electrical metering screen.

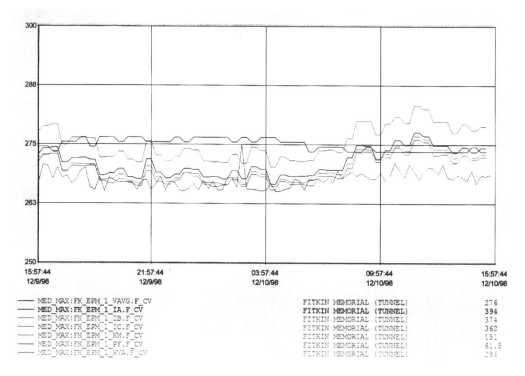

FIGURE 14.21   Example of an electrical trend graph.

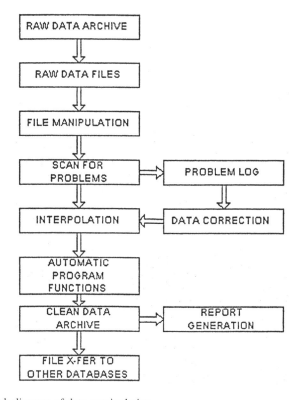

FIGURE 14.22   Block diagram of data manipulation.

Since utilities metering is an afterthought for all existing buildings, some buildings have more than one feeder, and therefore have more than one meter. Other buildings are not metered directly, and building consumption is determined by calculation from other meters. For this reason, so-called "building reports" are generated for each building monthly (Figure 14.23, see next page). The engineer responsible for report generation reviews and "cleans up" the meter reports, using available software tools as well as the inputs from engineers reviewing the metered data. This two-phase "data cleaning" process provides statistically valid and accurate data for the reports. Clean data are compiled for each building and the monthly totals are electronically transferred to the purchased utilities system. Here, the reported consumption is assigned to the respective accounts representing charges for buildings and/or departments.

The so-called residuals, which is the difference between the utility company's monthly readings and the totals reported by our metering system, has two origins:

1. Internal (line losses and metering errors).
2. External (due to a different monthly schedule of the utility company's meter reader — which is never at midnight of the last day of the month; therefore it is never in synch with the monthly reports generated by the computer).

The reports are reconciled monthly by allocating a prorated amount of the residual kWh equal to the consumption of the particular building. The discrepancy due to the utility meter reader's schedule is equalized over a billing season. The final number transferred to the purchased utilities system includes the metered consumption plus the calculated residual and does not require further data manipulation.

Other Reports

Since the stored data are in a spreadsheet format, it is easy to manipulate and create custom reports. Such custom reports may compare energy consumption of one building to another, monthly consumption to consumption of previous years, and other reports as needed by engineers and managers.

Reports can be presented in a graphical format, such as the one is Figure 14.24, (see page 223). The graph compares monthly consumption of a building over several consecutive years after an energy retrofit to the monthly consumption averaged over the 4 years prior to the retrofit. Some key points affecting consumption are indicated on the graph.

Electrical Systems Monitoring and Diagnostics

The Yale electrical distribution system is a complex web consisting of a cogeneration plant, the utility company's incoming electrical vault, normal and alternative distribution cabling throughout the campus, transformers and electrical switch-gear in the buildings. The system utilizes over 300 electrical meters throughout the campus. The metering system is also being used for diagnostics of the electrical system.

Prior to implementation of the metering system, power failures were reported to the University Control Center by building occupants affected by the failure. In case of a power outage that affected more than one building (failures of high voltage cables, fuses, or breakers), the extent and the origin of the failure were determined only from the calls originating from the affected locations. A traditional, low-tech approach to system monitoring and troubleshooting. While this approach was workable during regular working hours, off-hour and weekend electrical outages were a much greater challenge.

Development of the cogeneration plant and improvements of the distribution system made the electrical system more reliable, but also more complex. The metering system — as originally set up — provided a base for electrical distribution system diagnostics.

Since the CBUMS resides on the campus real-time network (Maxnet), it is accessible from several locations throughout the University, including the Control Center and the electrical supervisor's office.

Electrical Monthly Report - Science Buildings

	122: Seeley Mudd Lib. (SO_EPM_3)		123: 140 Prospect (14_EPM_1, 14_EPM_2, 14_EPM_3, 14_EPM_4)		124: 80 Sachem (14_EPM_2)		125: 70 Sachem (14_EPM_3)		126: 124 Prospect (14_EPM_4)		127: Ingalls Rink (IR_EPM_1)		128: Greeley Lab (GR_EPM_1)	
	KWD	KWH	KWD	KWH	KWD	KWH	KWD	KWH	KWD	KWH	KWD	KWH	KWD	KWH
11/01/98	50	696	46	815	8	103	10	134	5	70	255	4068	68	1281
11/02/98	96	1002	95	1402	8	116	16	183	6	83	245	4007	80	1468
11/03/98	105	925	104	1403	11	125	14	202	6	85	206	3834	81	1474
11/04/98	103	935	106	1406	9	129	14	201	6	91	191	3736	81	1470
11/05/98	98	1262	88	1441	8	131	15	200	6	86	241	4210	80	1532
11/06/98	113	1411	92	1261	7	112	16	213	5	83	246	4100	77	1449
11/07/98	188	1279	49	899	7	102	11	142	5	71	261	4276	69	1343
11/08/98	51	1054	53	937	7	109	11	147	4	70	229	4213	70	1353
11/09/98	119	1490	98	1500	12	140	15	210	6	94	232	3914	81	1502
11/10/98	100	1452	98	1395	10	136	15	231	5	86	258	3923	83	1533
11/11/98	187	1482	96	1486	9	116	13	183	5	88	246	4107	80	1475
11/12/98	105	1386	96	1422	9	110	13	179	6	81	217	3927	81	1485
11/13/98	89	1330	86	1223	9	116	13	176	6	79	187	3802	82	1465
11/14/98	79	1235	63	1000	8	102	11	137	5	59	198	3908	67	1375
11/15/98	59	1182	48	871	7	99	10	131	4	59	203	4096	67	1325
11/16/98	84	1481	96	1421	7	110	15	205	6	94	249	4083	86	1436
11/17/98	92	1355	95	1455	8	114	14	220	7	101	218	4030	81	1495
11/18/98	112	1494	90	1432	9	118	18	218	7	94	248	4084	83	1511
11/19/98	114	1440	98	1399	10	120	20	240	7	92	217	3906	83	1511
11/20/98	111	1456	89	1259	8	100	12	181	5	83	319	4913	81	1449
11/21/98	52	1079	29	676	7	94	13	160	3	66	280	4953	71	1335
11/22/98	51	1083	28	679	6	95	12	175	3	63	220	5155	68	1327
11/23/98	187	1437	90	1249	9	99	14	189	5	79	273	4957	69	1409
11/24/98	114	1441	86	1221	8	101	14	183	5	74	251	4163	74	1380
11/25/98	88	1385	91	1206	9	109	14	169	5	66	239	4069	72	1376
11/26/98	55	1109	28	654	8	105	11	159	4	56	226	3438	61	1247
11/27/98	55	1121	26	669	7	108	12	160	4	60	241	4255	69	1271
11/28/98	53	1094	25	669	9	110	11	160	6	62	276	4708	62	1250
11/29/98	53	1089	51	875	8	107	11	150	4	69	289	5174	66	1320
11/30/98	105	1461	90	1419	10	123	14	179	6	85	258	5018	78	1437
AVG KWD:	96		74		8		13		5		241		75	
Peak KWD:	188		106		12		20		7		319		86	
AVG KWH:	1255		1158		112		181		78		4234		1409	
Total KWH:	37646		34744		3359		5417		2329		127027		42284	
Load Fact.:	0.547		0.649		0.555		0.561		0.618		0.733		0.783	
BC total KWH	41411		38218		3695		5959		2562		139730		46512	

FIGURE 14.23   Example of an electrical building monthly report.

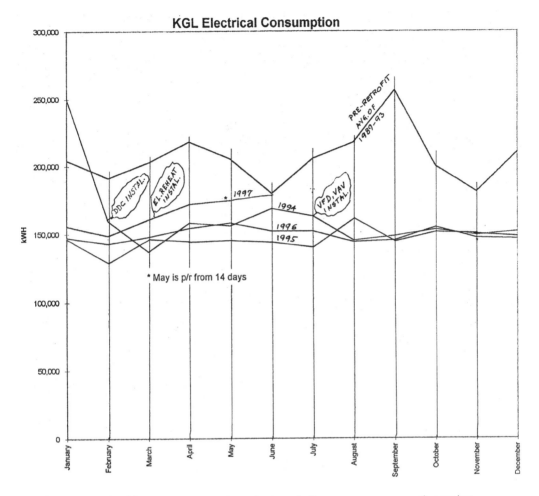

FIGURE 14.24 Monthly energy consumption prior to and after an energy conservation project.

From outside the University (let's say, home), the system is accessible via a Web page. Electrical engineers or supervisors can get online real time information (including reporting of alarms) on the electrical distribution system status and electrical values.

The metering system provides the following diagnostics for the operators and maintenance personnel:

1. Voltage and current unbalance of each metered point are continuously monitored. This serves as an early warning for the maintenance personnel on possible phase overloading.
2. Substantial voltage drop and drop of phase currents (i.e., to near zero) is reported as an alarm for each meter. When power failure occurs, an alarm message indicates its location. The meter(s) associated with the troubled spot would show zero values in the related dialog box, and would also change color from green to red on the color-graphic screen.
3. The power interruption is also visible and can be viewed historically on trend graphs, and consequently on the reports.
4. A separate screen is provided to show the values of exported electricity from the cogeneration plant as well as the amount imported by the plant from the utility. This screen indicates which feeders are live (have voltage and amperage readings) and/or have failed.
5. There is also a screen depicting high voltage normal and alternate service-switching schemes, which shows which feeders are live (hot) and which are stand-by.

Using the above tools, the Control Center personnel, the electrical supervisors and engineers, can determine the nature and exact location of electrical failures. Full diagnostic capabilities of the system have changed the entire approach to diagnostics and electrical system maintenance. Online diagnostics of system problems from within or outside of the University have shortened response and repair time and saved maintenance money, especially during off-hour call-back.

CHILL WATER METERING

Flow Computers without Communication to the Metering Server

Chill water metering at Yale is by ultrasonic, clamp-on meters. They were selected for ease of installation, good turn-down ratio, their diagnostic features, and communication options. In the earlier installations in the central and science areas of the campus, the meters' flow computers were connected to field panels, which were connected to the campus network via serial communications and Modbus communication drivers (Figure 14.25). This option was selected because none of the meters considered for installation (made by Panametrics and Controlotron) had "industry standard" communication options at that time. Due to lack of standard communications, only flow values were interfaced to the field panels via a 4- to 20-mA connection. Consequently, one could not take advantage of the flow computer features, such as diagnostics, online setup features, and other features.

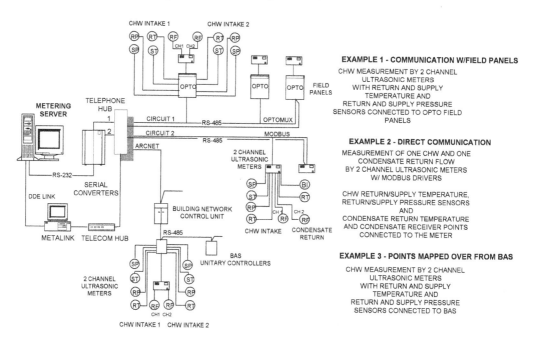

FIGURE 14.25 Flow computers connected to field panels.

The field panels at Yale are either Opto 22 panels connected via existing dedicated telephone lines using RS-485 serial communication and Optomux drivers to the CBUMS server, or Metasys DX controllers. DX controllers reside on the L2 (serial) bus connected to network controllers, which are on the campus-wide ARCNET. Metasys is interfaced to the CBUMS server via DDE interface called Metalink.

Besides flow from the flow computers, supply and return pressure and temperature sensors are also connected to the field panels.

The drawback of using Optos for data gathering is in the reduced accuracy of energy calculations and lack of CPU and memory in the panels. This causes unsynchronized timing of readings that are part of the same equation. While the flow computer scans the flow sensor at a fast rate (in seconds), the server's scan rate is much slower (in minutes). This introduces an error in the tonnage calculation, since the individual values used in the equation are read at different times (flow in seconds by the flow computer; temperature in minutes by the server).

The Optos do not have CPUs, they can't provide energy calculations, pulse totalization, data manipulation and data storage. Since all of these functions are in the server, this method of data acquisition requires a high degree of reliability of all components of the metering and communication system.

The calculation error is somewhat reduced by connecting the metered points to the Metasys DX panels. The DX panels also provide calculations, totalization, and faster scan rates than the metering system's server. Metasys can store "unlimited" amount of data, which can be uploaded to the server at regular intervals and preserved in both locations in case of failure of network components.

Another advantage of using DX controllers is that the Metasys — used primarily for building automation — is already in most buildings to be metered and is very reliable, with a minimum failure rate. The use of DX connections saved substantial cost for installation and system set-up. As said above, metering points from Metasys are mapped over to the Fix server using a Metasys DDE software link called Metalink.

Flow Computers with Modbus Communications

The communications shortcomings were eliminated during installation of the medical school metering system due to the willingness of Controlotron, Inc., to develop a Modbus interface for their meters. Controlotron meters with Modbus protocol option interface directly with the CBUMS server via RS-485 serial communications and Modbus drivers (Figure 14.26, see next page).

In addition to flow transducers, pressure and temperature sensors are also connected to the flow computers. Individual flow computers scan the connected field points and provide energy calculations (*chill water tonnage = flow × DT × .042*). Measured, calculated and diagnostic data along with the meter set-up parameters are stored in the flow computer registers and are accessible over the network using the Modbus addressing.

The meters are two channel meters. Their registers are set up so that each channel can be used for either chill water or condensate metering independently. The flow computers have sufficient number of inputs (4 to 20 mA, RTD, binary), and their registers are set up universally for the use of chill water and/or condensate metering for each channel.

Since the meters can be used for either chill water or condensate metering in any combination, engineers have to decide on use of one or two channel meters for a given location. This feature reduces not only engineering and set-up cost but also optimizes the number of meters used, thus reducing the installed cost (Figure 14.27, see next page).

Survey and Design

The chill water distribution system at Yale is an all-primary system, which means there are no secondary pumps, bridges, and other elements in the chill water system. Chill water pumped from the plant at a certain pressure is distributed throughout the campus into individual buildings. Building risers are "tapped" from the main distribution pipes. In most cases, return or supply flows, return and supply pressure, and temperatures are measured at each building riser. In some cases, the main supply temperature is measured at the main. Even though this is not as accurate as measuring the supply temperature for each building, it still provides sufficient accuracy since the chill water supply temperature is nearly constant throughout the distribution system.

The main parameter that moves the chill water through the coils is the pressure and the differential pressure. Since Yale has an all-primary system, it was prudent to check building pressure conditions at each location. Pressure profiles were developed at each building, consisting of static and forecasted dynamic pressure. Each building and associated cooling coils were checked to see

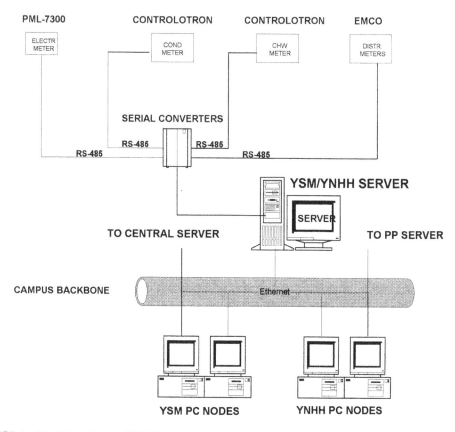

FIGURE 14.26 Example of a YSM/YNHH metering.

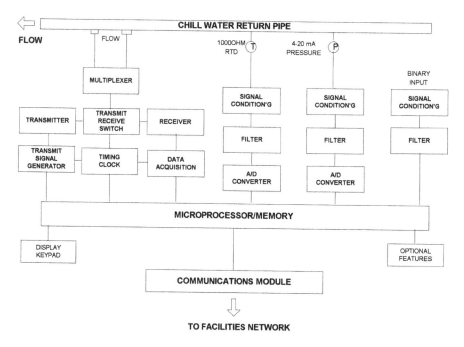

FIGURE 14.27 Block diagram of an ultrasonic energy meter.

if they were properly sized for the pressure conditions. Since the buildings were built before the conversion of the chill water distribution system to all-primary system (and certainly long before the chill water metering system installation), pressure conditions at each location were important. Despite the initial survey and the follow-up computer simulation, full understanding of the pressure profile was possible only after installation of the metering system.

Unique building pressure conditions exhibited themselves in buildings closest and farthest from the plant. The buildings closest to the plant have the highest supply and the lowest return pressure, and in many instances the differential pressure was high enough to "lift" the control valves of installed fan coil units. It was quite a surprise to find unintentional cooling in these buildings in the middle of December. The problem was solved with a unique design of DDC-controlled pressure reducing stations, similar to the ones used for the steam distribution systems. In all cases, the metering and the pressure reduction became integrated into the DX controllers. The high-rise buildings farthest from the plant were outfitted with flat-plate heat exchanger and secondary pumps. Isolating these buildings from the system allowed for reasonable pressure conditions throughout the distribution system.

Survey of the chill water distribution system along with installation of pressure reducing stations had enormous impact on the quality of chill water provided to the buildings.

Installation and Startup

Even though "clamp-on" ultrasonic meters were used, installation of the pressure and temperature sensors required installation of "hot taps" and "wells" for pressure and temperature sensors. Temperature sensors were matched pair RTD sensors for higher accuracy, since their reading is used and is crucial for accuracy of energy calculations.

Electrically, the metering system is connected either to Opto panels, Metasys DX controllers, or directly to flow computers as described in the previous sections. Data transfer to the server is via RS-485 communication and Modbus drivers.

System startup was a multifold process. First, the sensor signals had to be verified at the flow computer and field panel locations; second, the flow computers had to be set up and calibrated for each particular flow condition. Checking out of the communication between the server and the local devices was done only after having "credible" readings locally. Chill water distribution system parameters were verified after receiving data from each meter over the network.

Data Presentation

Chill water data (flow, temperatures, and pressures) are presented on color-graphic screens along with calculated tons for each building (consumption in tons of chill water is the measure used by the power plant, so it was decided to use the same units for ease of comparing production with consumption). The values are also available in trend graph formats.

In the medical school area, flow computers communicating to the server via the Modbus protocol have also other data displayed on the screen, similar to the data from the electrical meters. Upon "clicking" on the meter's dialog box, a set of pressure, temperature, flow values along with energy used, alarm, and diagnostic data appear on the right field of each screen (Figures 14.28 and 14.29, see next page).

Real-time data (color-graphic and trend graphs) are used by engineers evaluating systems performance, by power plant operators and managers evaluating export data (and making corrections for optimum production), and for system troubleshooting.

Report Generation

Chill water reports are similar to electrical reports (Figure 14.30, see page 229). The difference is that the chill water data are totalized in 1-hour intervals, since the 15-min peak demand is irrelevant

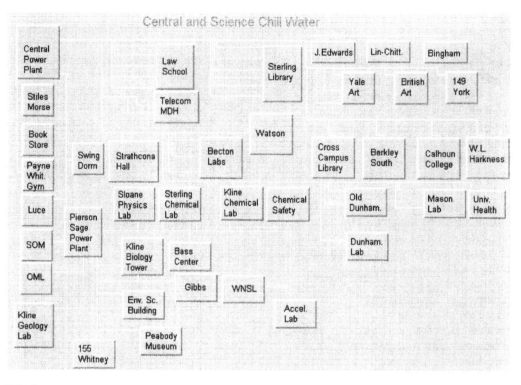

FIGURE 14.28   Chill water distribution overview screen.

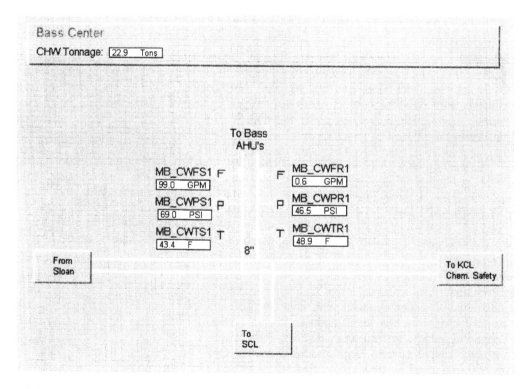

FIGURE 14.29   Example of a chill water color-graphic screen.

CHW Daily - Central Area

Time	SS_CWFS1 Loop GPM	SS_CWFS2 Strathcona GPM	SS_CWTD2 Strathcona F	SS_CWTN1 Strathcona Tons	BL_CWFR1 Becton Lab GPM	BL_CWTD1 Becton Lab F	BL_CWFR2 Becton Lab GPM	BL_CWTD2 Becton Lab F	BL_CWFR3 Becton Lab GPM	BL_CWTD3 Becton Lab F	BL_CWTN1 Becton Lab Tons	WA_CWFR1 Watson Lab GPM	WA_CWTD1 Watson Lab F	WA_CWTN1 Watson Lab Tons
0:00	1914.00	205.10	2.20	16.40	890.30	5.10	164.80	3.80	4.30	28.20	203.70	175.70	2.50	11.80
1:00	1907.20	206.10	2.20	19.70	877.40	5.10	162.60	4.80	3.30	28.20	198.00	175.70	2.50	11.80
2:00	1852.00	207.60	2.20	17.70	807.60	3.80	166.30	4.80	1.00	28.20	171.70	176.60	2.50	11.80
3:00	1831.50	217.80	2.20	19.10	777.40	5.00	174.50	3.80	1.00	28.20	194.80	182.30	1.30	9.80
4:00	1790.00	219.30	2.20	17.10	743.40	5.00	159.20	5.10	1.00	28.20	192.80	175.70	1.30	9.80
5:00	1751.90	214.90	2.20	17.90	734.40	5.00	174.00	3.90	1.00	28.20	187.40	176.90	1.30	9.80
6:00	1756.30	216.40	2.20	19.20	705.60	5.00	176.30	3.80	1.00	28.20	180.70	179.90	1.30	9.80
7:00	1733.80	215.40	2.20	26.20	715.60	5.00	164.80	4.90	1.00	28.20	181.30	172.40	1.30	7.20
8:00	1839.30	212.00	2.20	22.70	805.60	6.20	166.70	3.90	70.90	7.20	247.70	167.40	1.30	8.40
9:00	1843.70	212.50	2.20	19.80	794.70	6.20	155.80	3.90	75.60	7.80	255.20	165.90	1.30	10.60
10:00	1899.90	308.70	2.20	32.10	828.10	6.20	163.10	3.90	82.30	9.60	270.70	163.90	1.30	13.00
11:00	1898.40	215.90	2.20	24.40	850.10	6.20	158.00	5.20	84.90	11.00	261.30	167.20	1.30	14.10
12:00	1922.30	212.50	2.20	23.00	857.10	6.20	159.30	4.10	60.70	8.00	246.00	167.30	1.30	12.40
13:00	1898.40	209.50	2.20	21.20	844.50	5.10	152.90	5.10	62.50	9.40	233.70	170.90	1.30	14.00
14:00	1944.30	311.10	2.20	41.10	841.00	5.10	153.50	4.00	66.20	11.10	269.80	167.40	1.30	15.60
15:00	1956.00	311.10	3.30	43.20	843.70	5.10	156.30	4.00	58.90	8.30	250.80	171.30	1.30	14.70
16:00	1867.20	217.80	3.30	25.70	849.20	5.10	160.90	5.10	1.00	26.80	202.70	182.70	2.40	23.20
17:00	1863.20	215.90	3.30	24.00	871.50	5.10	160.40	5.10	1.00	28.20	199.70	177.40	2.40	14.90
18:00	1874.00	212.90	3.30	22.80	878.20	5.10	156.50	4.10	1.00	28.20	202.60	178.10	2.40	14.90
19:00	1873.50	212.00	3.30	22.80	879.10	5.10	152.90	4.10	1.00	28.20	204.40	178.10	2.40	15.70
20:00	1865.70	212.00	2.30	21.40	857.10	5.10	162.40	4.10	1.00	28.20	202.00	179.50	2.40	12.20
21:00	1810.50	220.80	2.30	25.70	781.00	5.10	167.40	5.20	1.00	28.20	203.00	181.40	1.40	10.90
22:00	1775.30	213.90	2.30	19.20	774.20	5.10	159.70	3.80	1.00	28.20	199.40	179.00	1.40	9.90
23:00	1781.20	215.40	2.30	20.90	768.90	5.10	163.00	3.80	1.00	28.20	204.10	179.50	1.40	11.60
0:00	1789.00	213.90	2.30	20.80	760.10	5.10	171.80	4.80	3.80	28.20	204.00	174.60	1.40	9.70
Max	1956.0	311.1	3.3	43.2	890.3	6.2	176.3	5.2	84.9	28.2	270.7	182.7	2.5	23.2
Min	1733.8	205.1	2.2	16.4	705.6	3.8	152.9	3.8	1.0	7.2	171.7	163.9	1.3	7.2
Avg	1852.1	225.7	2.4	23.5	815.7	5.3	162.1	4.3	24.3	21.8	215.1	174.7	1.7	12.4
Total	44449.6	5416.6	58.7	563.3	19575.7	126.1	3891.3	104.3	583.6	522.2	5163.5	4192.2	40.6	297.9

FIGURE 14.30 Example of a daily chill water report with loop GPM, individual building flows, differential temperatures, and calculated tons.

for power plant operation. The meter report shows flow totals, flow rates for each hour, and the related calculated tons. Daily reports are evaluated by engineers responsible for operation of individual areas of the campus, reviewed and compiled into monthly reports. Reconciliation of data from the export meters with the data from the building meters is done monthly and the residual is distributed proportionally to individual building consumption.

There are also special management reports which provide a quick overview of plant, distribution system, and building performance. These reports are printed out and distributed (during the cooling season) daily to the managers for performance evaluation.

Chill Water Distribution System Diagnostics

Online chill water diagnostics are used throughout the year, but mostly during cooling seasons. In the past, facilities engineers had to go to individual buildings to verify chill water readings from related gauges (temperatures and pressures). Deduction of chill water flow from differential pressures was more or less an educated guess.

Online diagnostic feature of the CBUMS system is a "dream come true" for any facilities engineer. It provides online flow measurement for each riser as well as measurements of pressures and temperatures, and calculations of their differentials. We have designed two sets of diagnostic tools (besides the real-time color-graphic screens, trend graphs, and hourly reports):

1. Diagnostic screens provide:
    a. Essential readings of "end" buildings of each branch
    b. Essential readings of every building on each branch line
    c. Bar charts of supply return and differential pressures throughout the distribution system divided into branches — probably the most visual presentation of the pressure profile of the entire distribution system
2. Daily reports provide hourly readings of building flow, differential pressure and temperatures, and energy consumption. The report is sectioned for each branch leg of the distribution system and provides total values of each branch and compares the branch values with the power plant export values. This report is distributed to O&M engineers and managers daily during the cooling season.

Both the diagnostic screen and the daily reports give the operating engineers good understanding of building operation. They also provide data for power plant optimization: optimization of pumping horsepower, chill water supply temperature, scheduling of equipment, and other. Managers benefit from clear understanding of operating conditions and building parameters and save on maintenance costs due to quick and pointed response to chill water problems.

Condensate Metering

Why Ultrasonic Condensate Metering?

One of the operating requirements imposed on installation of utilities metering is to limit shutdowns to a minimum and eliminating installation of inline meters for most applications (since they have to be cut into existing piping requiring prolonged shutdowns). Another factor that played a role in the decision to meter condensate (rather than steam) with ultrasonic meters was the limited selection of insertion-type flow meters. In fact, we did experiment with insertion-type vortex shedding steam meters with unsatisfactory results.

Based on the success with ultrasonic meters for chill water, the same meters were used for condensate metering.

Condensate from individual heat exchangers, coils, drip legs, etc., is collected into condensate receivers and/or liquid movers and then pumped back to the power plants. Since every building connected to the steam line (and condensate return) is measured, steam production is equal to the sum of individual building consumption plus residuals or losses. The distribution system losses are added (monthly) to individual building consumption proportionally to consumption.

Conversion of Condensate to Steam

The measured condensate flow (in GPM) is converted to steam (pounds of steam) by using a multiplier of 8.1 for a 180°F condensate temperature (GAL of COND × 8.1 = lb of COND = lb of STM). The equation provides fairly good results and accuracy without needing expensive in-line temperature sensors at the condensate return lines. If more accurate calculation is desired and there is a temperature sensor in the condensate return line, the particular coefficient can be calculated and the values can be found in the steam tables, using the specific volume Vf [cfpp] for actual condensate temperatures. The conversion coefficient is 1000 to convert the pounds of steam to BTU for normal condensate temperature and enthalpy.

The Importance of Condensate Temperature for Metering and System Troubleshooting

It should be said that while improved accuracy of the calculation does not warrant installation of temperature sensors in the condensate return lines, temperature readings are invaluable for troubleshooting of heating systems. Most buildings at Yale have installed condensate temperature sensors. A short program generates an alarm whenever the condensate temperature is greater than 190°F for a predetermined time (i.e., 30 min). The alarm is an indication of a failed steam trap allowing live steam into the condensate.

The same temperature sensors flag presence of steam in metered condensate pipes, which causes ultrasonic signal deterioration and inaccurate readings. Whenever the flag appears, the program disregards the meter reading (as unreliable), and interpolates the flow from two adjoining reliable readings.

Ultrasonic meter manufacturers provide tables indicating the relationship of the ultrasonic signal and the metered media temperature, which can also be used for alarm reporting.

Substitute Condensate Metering

Another method for condensate metering is counting the pulses from condensate receivers and liquid movers as events at pump-down. Since each event is equal to a number of gallons in the tank being pumped down, totalizing the pulses over time (an hour) provides hourly totals used for display and in reports. Conversion to pounds of steam and BTU is similar to the methods described above.

Although this method is not as accurate as direct metering by ultrasonic flow meters, the accuracy is sufficient for both customer billing and system monitoring. To improve the accuracy, the condensate return in the power plants and in individual branches of the distribution system is metered by ultrasonic flow meters. The calculated residual (the difference between the values provided by ultrasonic branch meters and the totals from the events of each branch building) is distributed to the metered buildings on each branch line.

Additionally, the event is also used for systems diagnostics. Using a time program function, we monitor the duration of each event. If the event is longer than a predetermined time (i.e., 5 min), an alarm is reported which indicates the malfunction of the medium pressure valve (in liquid movers) and/or float switch or relay of the condensate pump.

Communications to Condensate Meters

Flow computers of ultrasonic flow meters communicate with the CBUMS server via RS-485 serial communication with Modbus driver. Condensate temperature sensor as well as binary inputs of events are connected to each flow computer. The flow computer provides hourly totals as well as energy calculations, which are read by the server at scan intervals.

In instances when the condensate is calculated from the event, the binary input of the relay is connected to either Metasys DX-9100 controllers or Opto panels, along with the RTD temperature sensors. The server scans real-time values from field panels connected to the network, and provides all calculations (totals and energy).

This metering method does not provide for set-up variables, pipe sizes, and other variables provided by ultrasonic meters online.

PERSON MACHINE INTERFACE (PMI) — GRAPHICAL SCREENS, TREND GRAPHS, HISTORY FILES, REPORT GENERATION

PMIs are set up in the same way as for chill water metering. However, one has to understand the major difference between chill water and condensate metering. While chill water metering is based on continuous flow in the pipes, condensate metering is an interval metering — there is flow only while the vessel is being emptied during the condensate discharge cycle. During that time, the flow is constant, depending on pump characteristics or on the characteristics of the liquid mover. Therefore, it is more prudent to display on the screen the condensate flow of the last hour rather than instantaneous flow in gpm, since the gpm is the same whenever the receiver is being discharged.

MAXNET AND ITS USE

Facilities management cannot exist without information. One of the most valuable pieces of information is the real-time information provided by Maxnet over the network to facilities engineers, managers, and building occupants.

CONTROLS, AUTOMATION, AND INFORMATION ON THE LOW END

Real-time controls and automation systems control the power plants and the buildings. The systems are monitored by the power plant and by control center operators.

Individual C&A systems in the power plants control, monitor, and optimize the operation of the power plants. At this level, systems from different vendors are interfaced to the DCS of each plant. The DCS provides uniform operator interface, optimization, and reporting functions. The communication interfaces at this level are either serial communication or higher level Ethernet connections. The interfaces are via industry standard communication drivers.

Building automation systems control and operate the respective buildings. They were implemented over two decades of automation at Yale. Over the last few years they were migrated into the Metasys system, providing a common interface to the operators. The communication at the building level is vendor specific due to the policy of having each building controlled by a specific DDC system. Fume hoods, VAV drives, and other third-party systems in the buildings are interfaced to the BAS system via individual communication drivers. The new requirement to provide LonMark® communication at the building level is to unify the communication protocols in a multivendor environment.

The Central Building Utilities Metering System provides information to facilities engineers and managers. The meters communicate to the server via serial communication and Modbus drivers.

The systems in all three geographical areas of the University are networked on the Maxnet and the information is available either on the network or on the web (Figure 14.31).

DISTRIBUTION OF INFORMATION

The information from the three areas (production, distribution and building operation) is integrated via the Maxnet servers (Figure 14.32, see page 234). The servers reside on the campus Ethernet

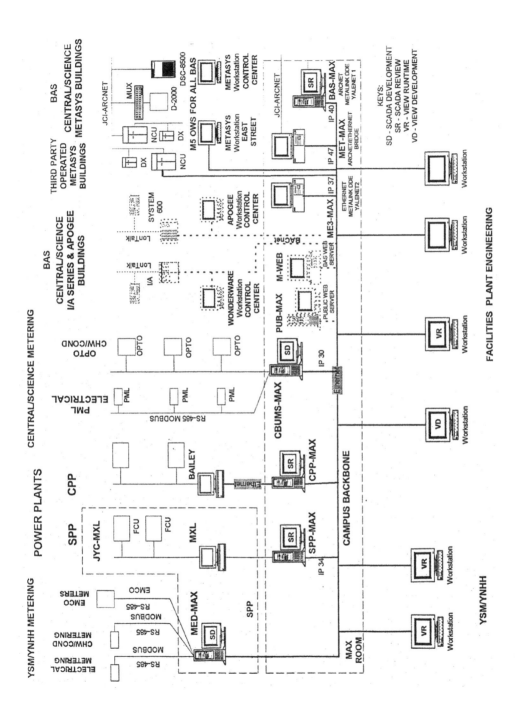

FIGURE 14.31 Yale Maxnet.

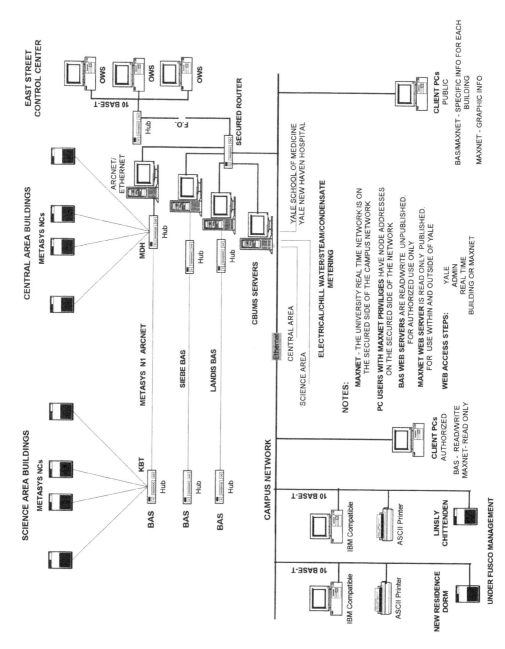

FIGURE 14.32 Information distribution on Yale Maxnet.

and scan the data from their respective automation or metering systems in the three areas of the University. The systems are polled at 2- to 4-min scan rates. The users can access real-time information residing on the servers in three different ways:

1. Via the University network — Facilities engineers and managers who are involved in daily management of their assigned areas have direct access to data on the servers from their office computers. To be able to get access, they must have a software license (a key), just as do the operators in the power plants and control center.
2. Via so-called facilities web servers — This important feature provides access (no license needed) for supervisors, zone managers, facilities engineers, and managers. Since these key people are located in buildings throughout the University, access without using web servers would require substantial investment into networking and software licenses. They can access the web from anywhere (i.e., from home) providing a useful troubleshooting tool, especially for off-hour and weekend operation. This is probably the least costly and the most useful feature of the network.
3. Via public web server — The University decided to provide essential information on a so-called public web server to the University community. The public server can be accessed through Yale Web site from anywhere within or outside the University. The information is available to customers, such as students, faculty, staff, business managers, as well as anyone who is interested in looking at the building data. Customers can select the building they are interested in from the Web page. The public Web server accesses the information from the building automation, metering, and power plant servers, and compiles it in a format displayed on the public server screens.

The above network provides pertinent information to users, providing sufficient security to maintain system integrity without expensive and complicated system security measures. Network and system security issues were omitted from this discussion.

COST ALLOCATION AND BENEFITS

To provide a total cost for the entire installed system would be a difficult task, since the system was implemented over several decades. However, Yale policy was and still is to preserve previous investment into the existing systems. An evidence is the Delta 2000 field panels installed in the late 1970s, upgraded to EMS in the early 1980s, interfaced again to Pegasus in the late 1980s and migrated into Metasys in 1999.

Nevertheless, networking and systems integration is an investment that is difficult to justify up-front on returns from the operating budget (manpower, equipment, and energy). The benefits demonstrate themselves as improved operation, savings on labor cost due to diagnostics, avoidance of failures due to advance warning, energy savings due to optimum utilization of the resources and consequent reduction of production costs, and savings due to availability of information.

To see the incremental cost of integration, we did a cost allocation breakdown during the last phase of the metering system installation (see Figure 14.33, see next page). The breakdown shows that about 87% of the total cost is for field installations. This may be slightly higher than for previous installations, due to dedicated wiring (in conduits) that had to be provided as a result of inadequate telecommunication wiring. (Note that existing telecommunications wiring was used for communications from the meters to the servers).

About 13% of the cost is for communications, drivers, and applications software work on the servers and client workstations. This suggests that networking and distribution of information to clients (and by clients, is meant just about anyone interested in the information, since the information is available on the web) is a fraction of the total installation cost. In other words, for this incremental cost the information that otherwise would not be available is distributed to clients who previously

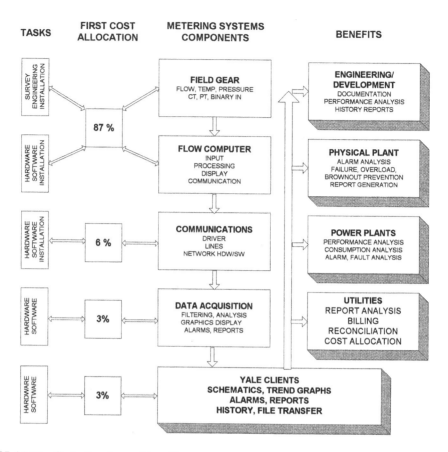

FIGURE 14.33 Cost allocation and benefits.

did not have access to such information. The cost for integration elevates facilities management to a completely different platform, providing information, analytical tools, and reports that were never available for problem analysis and maintenance. There is more information available for system review, such as operating parameters, which makes the operation less prone to errors.

The combination of graphical screens, trend graphs, and reports provides sufficient details to clients analyzing system or building performance to do a thorough and professional job.

The most significant benefits the system provides to the University can be divided into the following categories:

- **Metering of utilities for billing and cost allocation.** Monthly reports provide clean data of energy consumption for each building. The reports are distributed to managers and engineers for performance evaluation. The clean data are electronically transferred to the purchased utilities system, which provides energy accounting and cost allocation to individual buildings or departments, an important function similar to the billing function of a utility company. Proper and credible cost allocation is required for justification of research costs, adding credibility to reporting and allowing the University more accurately to allocate cost of energy consumed in research.
- **Optimization of power plant production based on the actual consumption data.** While power plant optimization is a function associated with individual power plant controls and monitoring systems, meeting the demand in a timely fashion is made possible by monitoring the distribution systems parameters. For example, chill water pumping

horsepower is reduced to meet the actual pressure profile of the entire distribution system. Chill water demand during off-peak season is being met with increased chill water temperature, thus saving on chill water production cost. Because the operators have the information from production, buildings, and distribution, they can make the proper decisions.

- **Alarm reporting of the distribution system parameters, buildings, and power plants** provides an add-on layer to monitoring. Facilities engineers can monitor the systems and its parameters directly from their office computers.
- **System diagnostics.** This feature saves probably the most maintenance dollars due to online diagnostics of systems parameters from remote locations — from the offices of managers, engineers, and supervisors. The system provides online diagnostics before a technician or mechanic is sent to the troubled spot. Supervisors can send the right person with the right tools with the right information directly to the problem spot. This is a major improvement from waiting until the problem is reported by the occupants, then sending a mechanic to analyze the problem, and only then sending the right crew to fix the problem. Now managers can respond to a problem before it has an impact on the system or building operation and in many instances before it inconveniences building occupants.
- **Energy savings** due to customer awareness, optimization of power plant and building parameters, modification and redesign of systems, minimizing cost for make-up water, chemicals in the power plants, reduction of consumption and peak demand, etc.
- **Minimizing damage to experiments, equipment, and furnishings in the building** due to continuous monitoring, performance analysis, and early detection of problems.
- **Protection of investment into systems** by allowing future upgrades and interfaces to the existing networks for minimum cost, using industry standard components. Since the system was put together by participation of in-house engineers, there is already a trained work force to operate and maintain the system.

CONCLUSION

Even before full completion of Maxnet, the demand for information and for more features had increased. The popularity of the system and its recognition by upper management increased dramatically upon dissemination of the information via the web. Despite its complexity, the system is up and running without major "hick-ups" and without dedicated network management. The system's daily operation, modifications, and its management, just like its implementation, has been done by the dedicated effort of the Plant Engineering staff. Without this effort, the system would never have been developed to become a universal tool utilized by the entire University.

Bibliography

Amborn, R. and Ehrlich P., Decision steps for implementing a BACnet interface project, *ASHRAE J.*, November 1994.

ANSI/ASHRAE 114-1986: ASHRAE Standard, Energy management control systems instrumentation, ASHRAE, 1987.

ANSI/ASHRAE Standard 135-1995, BACnet, a data communication protocol for building automation and control networks, ASHRAE, 1995.

ASHRAE Technical Data Bull., Vol. 5, No. 5, DDC and Building automation systems, ASHRAE, 1988 and 1989.

Baird, L. J., York is an "open" company, Protocol Forum, *Heat./Pip./Air Cond.*, August 1994.

Bartree, C.T., Data Communications, Mnetworks and Systems, *SAMS*, 1991.

Bartunek, I., CAN—Controller Area Network, Automatizace, No. 4 and 40, 1996 and 1997, Czech Republic.

Bernaden, J. and Williams, A., *Open Protocols: Communication Standards for Building Automation Systems*, Fairmont Press, 1989.

Black, U., *TCP/IP & Related Protocols*, McGraw-Hill, New York, 1995.

Boed, V., Networking is key to university's building automation strategy, Technology Report, *Energy User News*, October 1993.

Boed, B., *Controls and Automation: Applications Engineering*, CRC Press, Boca Raton, FL, 1998.

Bowker, R., *Interoperability*, Racal InterLan, 1989.

Cilia, J., *Building and Facilities Automation Systems*, Fairmont Press, 1991.

Contemporary Control Systems, ARCNET, Factory LAN Primer, 1987.

Cornillaud, M., Open protocol: freedom to choose optimum controls, Protocol Forum, *Heat./Pip./Air Cond.*, August 1994.

Decotiguie, J.D.D. and Ruiz, L., FIP Fieldbus and real time distributed automation, Swiss Federal Institute of Technology, 1998.

EIA, EIA – 485 Standard, Electronic Industries Association, 1983.

Elyashiv, T., Beneath the surface: BACnet Data link and physical layer options, *ASHRAE J.*, November 1994.

Fedrizzi, S.R., Going green: the advent of better buildings, *ASHRAE J.*, December 1995.

Fieldbus Foundation, Foundation Fieldbus Technical Overview FD-043 Revision 1.0.

Freer, J.R., *Computer Communications and Networks*, IEEE Press, 1996.

Goldschmidt, G.I., A data communications introduction to BACnet, *ASHRAE J.*, November 1994.

Happ, H.R., The remote telephone interface, *Heat./Pip./Air Cond.*, April 1995.

Hartman. T., Direct digital controls fundamentals, *Heat./Pip./Air Cond.*, February 1995.

Hartman, B.T., *Direct Digital Controls for HVAC Systems*, McGraw-Hill, New York, 1993.

Hartman, T., Practical considerations for protocol standards, *Heat./Pip./Air Cond.*, August 1994.

Hetherington, W.T., The Canadian response to open systems for building automation, *ASHRAE J.*, November 1994.

Hrdlicka, M., Fieldbus Introduction, VUT Brno, Czech Republic, 1998.

Hull, G.G., BACnet: Expectations running too high?, Protocol Forum, *Heat./Pip./Air Cond.*, August 1994.

INTERBUS- S Network at the sensor actuator level, *Automatizace*, No. 1, 1996.

The International P-Net organization: The P-Net, Fieldbus for Process Automation, Czech Republic, 1995.

James, J., Landis & Gyr Powers sizes up BACnet, Protocol Forum, *Heat./Pip./Air Cond.*, August 1994.

JCI, How to Choose the Right Facilities Management System, Johnson Controls, Inc., 1993.

Johnson Controls, Inc., Bernaden, J. and Neubauer, *The Intelligent Building Sourcebook*, Fairmont Press, 1988.

Kammers, K.B., Beyond open protocols, *ASHRAE J.*, December 1994.

Kriesel, W.A. and Madelburg, O.W. Eds., ASI e—The Actuator-Sensor-Interface for Automation, Carl Hauser Verlag, Munich, 1995.

Landman, J.W., Linking HVAC system design and operation through integrated controls, *ASHRAE J.*, December 1995.

LoGalbo, V.R., Carrier stands behind standard protocol, Protocol Forum, *Heat./Pip./Air Cond.*, August 1994.

LonMark Foundation, Interoperability Guidelines for Layers 1–5 and for Application Layer, Echelon Corp., 1998.

Madan, P., Overview of Control Networking Technology, Technical Paper.

McGowan, J., *Networking for Building Automation and Control Systems*, Fairmont Press, 1992.

McGowan, J.J., *Direct Digital Control*, Fairmont Press, 1995.

Newman, H.M., *Direct Digital Control of Building Systems*, John Wiley & Sons, New York, 1994.

Nordeen, H., Fundamentals of control from a systems perspective, *Heat./Pip./Air Cond.*, August 1995.

Petze, J., Modularity and the design of building automation systems, *Heat./Pip./Air Cond.*, August 1995.

The Profitbus Standard, Profitbus International, Germany, 1993.

Public Works Canada, Canadian Automated Building Protocol, PWC, Ottawa, 1992.

Seyer, D.M., *RS-232 Made Easy: Connecting Computers, Printers, Terminals and Modems*, Prentice-Hall, Upper Saddle River, NJ, 1984.

Spragings, D.J., Hammond, L.J., and Pawlikowski, *Telecommunications Protocols and Design*, Addison Wesley, Reading, MA, 1992.

Suzukida, J., Trane supports ASHRAE BACnet protocol, Protocol Forum, *Heat./Pip./Air Cond.*, August 1994.

Tuft, D., Open systems: Honeywell's customer response, Protocol Forum, *Heat./Pip./Air Cond.*, August 1994.

Wahlquist, D., Siebe supports BACnet and LonWorks, Protocol Forum, *Heat./Pip./Air Cond.*, August 1994.

Warrior, J., HART—An open protocol for distributed sensor applications, *Sensor 95*, Kongressband Proc.

Weaver, R.T., Johnson incorporates open protocols and standards as they emerge, Protocol Forum, *Heat./Pip./Air Cond.*, August 1994.

Index